MRI ESSENTIALS FOR INNOVATIVE TECHNOLOGIES

MRI ESSENTIALS FOR INNOVATIVE TECHNOLOGIES

GIUSEPPE PLACIDI

CRC Press
Taylor & Francis Group
Boca Raton London New York

CRC Press is an imprint of the
Taylor & Francis Group, an **informa** business

CRC Press
Taylor & Francis Group
6000 Broken Sound Parkway NW, Suite 300
Boca Raton, FL 33487-2742

First issued in paperback 2019

© 2012 by Taylor & Francis Group, LLC
CRC Press is an imprint of Taylor & Francis Group, an Informa business

No claim to original U.S. Government works

ISBN-13: 978-1-4398-4040-5 (hbk)
ISBN-13: 978-0-367-38143-1 (pbk)

Visit the Taylor & Francis Web site at
http://www.taylorandfrancis.com

and the CRC Press Web site at
http://www.crcpress.com

To my family

The author's proceeds from the sale of this book will be entirely donated to Bambin Gesù Children's Hospital in Rome.

vi

Foreword

Magnetic resonance imaging (MRI) is the fastest-growing modality in diagnostic imaging. There are now thousands of MRI facilities around the world employing a huge number of specialists from different backgrounds.

This book is intended for those working in this area who need a more than superficial knowledge of emerging MRI techniques.

The book follows a logical sequence that starts with the mathematical basis required to understand magnetic resonance physics. The basic concepts of MRI and the principles of conventional MRI techniques are then illustrated in a concise manner. These early sections should serve as background for the more novice reader, and may be skipped by the more experienced. From this point, a series of more complex concepts, which form the primary topics of the book, are presented. These encapsulate several novel acquisition techniques that relax the need for a highly uniform static magnetic field, and others that can improve the performance of MRI when undersampling is applied.

The deep physical insight into the relationship between spin dynamics and gradient field modulation expressed in this book paves the way to a completely new family of pulse sequences that may represent a new line of development for MRI.

As a whole, this excellent book should be found on the bookshelves of all those specializing in the field of magnetic resonance.

Antonello Sotgiu,
Professor and Chairman
Department of Health Sciences
Università dell'Aquila
L'Aquila, Italy

Foreword

Magnetic resonance imaging is recognized to be one of the most important medical advances of the 20th century. It was developed thanks to the joint efforts of physicists, chemists, and engineers, and further improved by the contribution of MRI users, the physicians.

In fact, physicians always have increasing demands both in diagnosis and clinical applications, given the richness of parameters that open MRI to a large number of possible applications. These include high-resolution real-time imaging (regarding, for example, functional MRI or cardiac imaging), the substitution with MRI of invasive x-rays fluoroscopy in interventional procedures (such as vascular catheterization), the massive use of whole-body MRI for cancer prevention. In order to fulfill these requests, though, further technological improvements need to be made in areas including the design of extremely accessible (completely open) scanners, the implementation of real-time imaging sequences, the improvement of spatial resolution, the reduction of artifacts due to magnetic field in-homogeneity, and the design of real whole body scanners.

In this book, Dr. Placidi presents a series of state of the art acquisition sequences and algorithms that aim to address several of these technological challenges yet require little hardware modification. It represents the emergence of a very disruptive approach to MRI.

The author divided the book into four logical sections:

1. basic concepts and conventional techniques,

2. limitations to the advancement of medical application based on conventional MRI,

3. suggestion of innovative solutions,

4. discussion on the future of MRI in view of the use of the proposed advanced methods.

The book combines rigorous development of mathematical concepts with descriptive presentations of the ideas; covering a specialized area of MRI, yet most chapters are accessible to readers with various levels of expertise. It is complete, readable and usable both for research and for specialized learning purposes in areas of advanced MRI imaging techniques.

Dr. Placidi indicates a path useful to both physicists and engineers for implementation, and to radiologists and clinicians in thinking of possible new ways to apply the suggested methods, and so continues the process of invention and innovation in the still only partially explored field of MRI.

Roberto Passariello, MD
Professor and Chairman
Department of Radiological Sciences
Policlinico Umberto I
Università di Roma La Sapienza
Rome, Italy

Acknowledgments

I acknowledge and thank all of those who supported me in many ways during the completion of the project, in particular:
- Antonello Sotgiu and Roberto Passariello for their appreciation of my endeavour and supporting me with the kind words used in their forewords to this book
- Paul Summers who proofread the book and whose contributions improved it deeply
- Carmelita Marinelli for her valuable help in artworks preparation
- Marcello Alecci and Angelo Galante for their useful suggestions
- Springer, Elsevier, IEEE and Bentham Publishers for allowing the reproduction of some of my previously published papers at no cost
- Banca di Credito Cooperativo di Roma for having partially funded this project.

My deepest gratitude goes to my family for encouraging me graciously, and to my fraternal friends and colleagues Pasquale Gallo and Giancaterino Gualtieri (they know why).

Preface

Magnetic resonance imaging (MRI) is recognized as one of the most advanced computerized imaging modalities. This imaging technology provides capabilities to reveal structure and function of the human body with a level of detail beyond that obtained with other diagnostic or its unique flexibility.

The initial clinical installations were in major medical research centers around the world, including one in the Radiological Centre of the San Salvatore Hospital in L'Aquila, Italy, directed by Prof. Roberto Passariello.

Since these first installations some 30 years ago, systems have been widely commercialized worldwide. Nowadays, whole body installations are often complemented by dedicated smaller systems used for the study of extremities and can be transported where necessary.

This revolution in medical imaging originated from the work of Bloch at Stanford and Purcell at Harvard. Both reported their discoveries of nuclear magnetic resonance (NMR) independently, in 1946, in separated papers [11], [112]. The importance of their work earned Bloch and Purcell the Nobel Prize in physics in 1952.

Many researchers have developed techniques that have moved these discoveries from physics into different applications. Chemists discovered the *chemical shift* effect, that is, a small shift in the frequency of the NMR signal caused by the fact that the frequency of a particular nucleus depends also on the chemical environment where it is situated. The recording of chemical shift and other atomic interactions, known as NMR spectroscopy, helps in determining molecular structure.

Bloch made the first biological measurement about the time of the NMR discoveries, by placing his finger into his pioneering instrument with the subsequent production of a strong signal due to the spin of the hydrogen nucleus.

Thereafter, the construction of high field magnets with bores of several centimeters allowed whole perfused organs to be accommodated, opening the way to extensive studies in NMR spectroscopy of human tissues.

Advancing from spectroscopy to MRI, Lauterbur presented the first imaging result in 1973 [66], obtained by localizing NMR signals through the use of magnetic field gradients, and in 1974 he presented the first imaging sequence [67].

During the same period, Mansfield developed a mathematical model to analyse signals from within the human body in response to a strong magnetic field, as well as a method to achieve very fast imaging (echo-planar imaging) [77].

For their inventions, both Lauterbur and Mansfield were awarded the Nobel Prize in Physiology or Medicine in 2003.

The contribution of Damadian who found, in 1970, the basis for using MRI as a tool for medical diagnosis by noting differences in the MR signal between cancerous and normal tissue [26] was also fundamental.

For the first time in history a radio signal originating inside human body was used to monitor tissues properties in vivo from outside the human body.

In 1974, Damadian filed the world's first patent in the field of MRI issued by the U.S. Patent Office, entitled "Apparatus and Method for Detecting Cancer in Tissue" [27] and, in 1977, he completed construction of the first whole-body MRI scanner.

An early evolution in MRI applications was the passage from continuous wave radiofrequency (RF) excitation and static field gradients to the use of pulsed RF and switched field gradients to collect data; it was a revolution mainly due to the invention of innovative acquisition sequences.

Since then, a great number of innovations have advanced MRI in areas including main magnetic field design and technologies, gradients and RF coils design, acquisition sequences, and electronic devices.

Despite this progress, exciting challenges still await MRI, due to its enormous potential in applications ranging from real-time imaging, to open systems, to interventional imaging.

It is time for MRI to undergo another revolution; with a proper choice of smart encoding methods and acquisition sequences (either using a priori information about the sample or "blind" sparse sampling techniques and reconstruction) MRI can reach ambitious results with few or no hardware modifications.

These methods are the subject of the present book. Very specialized acquisition sequences are reported to indicate the direction for innovative MRI, in a speculative sense.

The book covers a restricted piece of the MRI world. For this reason it is not in competition with existing, fundamental, complete handbooks on MRI. For the same reason, it has been assumed that the readers have already acquired basic knowledge of mathematics and physics of MRI, though some fundamentals are treated in this book.

Here the progression obeys to the following rule: first the problem to be solved is described, then the proposed solutions. For some of the proposed solutions, being speculative and innovative, only numerical simulations are reported, as experimental data is very preliminary or unavailable. Some of the reported results have not yet been published elsewhere.

The treated arguments can be considered building blocks to be combined and refined to overcome current barriers in the field of MRI.

During the past 19 years, the author has been involved in these topics, having taken his first steps in the field of electron paramagnetic resonance imaging (EPRI), both continuous wave and pulsed wave, in one of the world's pioneering EPRI labs, that directed by Prof. Antonello Sotgiu, at the University of L'Aquila. The research in the field of EPRI, especially in pulsed EPRI, enabled the development of similar methods in the field of MRI.

The book is divided in chapters. Chapter 1 covers basic mathematical tools. Chapter 2 describes the physics of MRI, in a phenomenological sense, covering some basic conventional imaging sequences, the hardware composing a MRI scanner and, finally, a virtual MRI device for numerical simulation of imaging sequences. Chapter 3 describes the severe artifacts produced by conventional MRI with attempts to deal with advanced applications; residual magnetic field inhomogeneity and undersampling effects are discussed. Chapter 4 describes some conventional methods used to deal with residual magnetic field inhomogeneity and a completely different approach to the problem, based on time-varying gradients and temporal frequency variation coding (acceleration). Chapter 5 covers two innovative acquisition methods capable of reducing acquisition time, motion, and undersampling artifacts: adaptive acquisition and compressed sensing. These methods have the same purpose, but they use different philosophies to pursue it. Adaptive methods use the definition of information content to collect only the most informative data; they adapt to the shape of the analyzed sample. Compressed sensing uses general sampling patterns and hypotheses about the imaged samples, for example, the sparseness, independent of one specific sample shape. The chapter concludes by describing how these methods can be used in a mixed sense to test the possibility to retain the advantage of both, thus reducing the measuring time while improving image quality. In Chapter 6, some predictions for the future of MRI, regarding the innovative methods presented in this book, are discussed.

xvi

List of Figures

xx

Contents

Symbol description

δ	Impulse function, also called Dirac function
shah	Sum of equispaced δ functions, also called Dirac comb
$\otimes$	Convolution between functions
$\cdot$	Product point by point between functions
z^*	Conjugate of z
j	Imaginary unit
$\prod$	Nonperiodic square pulse
$sinc(x)$	Represents the function $\frac{sin(\pi x)}{\pi x}$
$acc_s(t)$	Acceleration $sinc$ function
ν	Frequency
ω	Angular frequency
$\boldsymbol{I}$	Nuclear spin
$\boldsymbol{\mu}$	Magnetic moment

γ	Gyromagnetic ratio
$\mathbf{B}_0$	Main magnetic field, supposed directed along the z direction
$\mathbf{M}$	Magnetization vector
$\mathbf{M}_{xy}$	Transversal magnetization component
$\mathbf{M}_z$	Longitudinal magnetization component
$\mathbf{B}_1$	Radiofrequency magnetic field
T_1	Longitudinal relaxation time
T_2	Transversal relaxation time
T_2^*	Transversal relaxation time affected by perturbing magnetic fields
$\mathbf{G}$	Magnetic field gradient
T	Tesla

Abbreviations

NMR	Nuclear magnetic resonance
MRI	Magnetic resonance imaging
$fMRI$	Functional magnetic resonance imaging
$EPRI$	Electron paramagnetic resonance imaging
FT	Fourier transform
FT^{-1}	Inverse Fourier transform
FBP	Filtered back projection
FR	Fourier reconstruction
FID	Free induction decay
FOV	Field of view
$rFOV$	Reduced field of view
VOI	Volume of interest
ROI	Region of interest
RF	Radio frequency pulses used to excite the sample
PW	Power spectrum
SNR	Signal to noise ratio
MSE	Mean square error
$PSNR$	Peak signal to noise ratio
PSF	Point spread function
CS	Compressed sensing
TR	Time of repetition
TE	Time of echo
TA	Time of acquisition (interval of acquisition)
IR	Inversion recovery
$1D$	One dimensional
$2D$	Two dimensional
$3D$	Three dimensional
ppm	Parts per million
dB	Decibel
FSE	Fast spin echo
$GRASE$	Gradient and spin echo
EPI	Echo planar imaging
MRA	Magnetic resonance angiography
$CE-MRA$	Contrast enhanced magnetic resonance angiography
$PROPELLER$	Periodically rotated overlapping parallel lines with enhanced reconstruction
$RIGR$	Reduced-encoding MR imaging by generalized series reconstruction
$PR\text{-}rFOV$	Projection reconstruction reduced field of view
$UNFOLD$	Unaliasing by Fourier-encoding the overlaps using the temporal dimension
$k\text{-}t\ BLAST$	$k\text{-}t$ Broad-use linear acquisition speed-up technique
$SMASH$	Simultaneous acquisition of spatial harmonics
$GRAPPA$	Generalized autocalibrating partially parallel acquisitions
$SENSE$	Sensitivity encoding
$BOLD$	Blood oxigen level-dependent
$SRS\text{-}FT$	Sparse radial scanning - Fourier transform
$POCS$	Projection onto convex set

Part I

Basic Concepts

2

1

Mathematical Tools

CONTENTS

This chapter introduces a selection of mathematical tools used throughout the book.

Most of the reported results are extensively treated in [12] and in [13]; detailed discussions and demonstrations can be found in the same references.

Frequency encoding is given particular attention, with consideration of its relevance to signal formation and decoding using Fourier transform. Fourier theory is used in many scientific fields as a mathematical or physical tool to modify a problem into one that can be more easily solved. In MRI, Fourier theory can be considered as a physical phenomenon, not simply as a mathematical tool. For these reasons, it will be treated in detail. In particular, the sampling theorem and the Nyquist's rule are introduced and discussed to highlight the sufficient conditions of sampling and to allow the introduction of smart undersampling techniques that are discussed in later chapters.

The chapter ends with the description of subjective and objective methods used to evaluate and compare processed images or signals.

1.1 Frequency Encoding and Fourier Transform

The key to understand MRI signal formation and processing is to think in the frequency domain. An essential step in doing so is to appreciate how a collection of signals with defined frequencies relates to a periodic function.

A periodic function is a function defined in an interval T_0, called its period, that repeats itself outside the interval. The sinusoid $\sin(x)$ is the simplest periodic function and has an interval of 2π.

It is easy to see that the sine function is periodic since $\sin(x + 2\pi) = \sin(x)$. The sine function $\sin(2\pi\nu x)$ repeats itself ν times in the interval from

0 to 2π. Its frequency is ν, and it is measured in cycles per second (or Hertz, abbreviated in Hz), for temporal signals (x is time), or cycles per millimeter for spatial objects (x is space).

Any periodic function can be written (decomposed) as a sum of sinusoidal waves, also called sinusoids (or harmonics), whose periods are integer divisors of the period of the original periodic function. In determining the sum, we are free to choose the coefficients of sinusoids, their frequency, and phase, and we can use an infinite number of higher and higher frequencies.

The resulting function is also periodic (since all the sinusoids are periodic). The above consideration assures that such a sum can always be found.

In this model, the frequencies of the sinusoids are related because they are multiples of a fundamental frequency. This means that, although the original function is continuous, its representation as a sum of sinusoids can be treated as a discrete (integer, i.e., "sparse"), simpler, function.

As an example, consider a periodic square pulse such that shown in Figure 1.1 whose amplitude oscillates between 0 and 1.

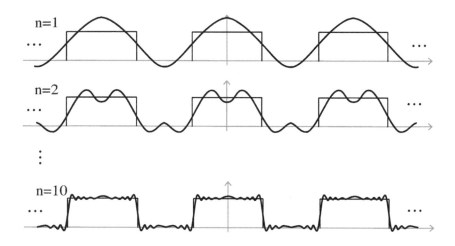

Figure 1.1
Approximation of a periodic square pulse (for simplicity, just three periods are shown) as a sum of sinusoids of different frequency and amplitude. By increasing the number of terms, the sum converges to the square pulse. Note the oscillations close to the edges, where the square wave is discontinuous; this effect is called Gibbs phenomenon.

This function is discontinuous at the edges of the pulse.

A square pulse can be written in the limit as $n \rightarrow \infty$ as the series

$$
\begin{aligned}
p_n(x) &= \frac{1}{2} + \frac{2}{\pi} \sum_{k=1}^{n} (-1)^{k-1} \frac{\cos{(2k-1)}\omega_0 x}{2k-1} \\
&= \frac{1}{2} + \frac{2}{\pi} (\cos \omega_0 x - \frac{1}{3} \cos 3\omega_0 x + \frac{1}{5} \cos 5\omega_0 x + \ldots) \quad (1.1)
\end{aligned}
$$

where the fundamental angular frequency (in radians) is $\omega_0 = 2\pi/T_0$.

Equation 1.1 represents the Fourier series of the periodic square wave. It shows the first terms of the approximation; note that just cosines terms are present. Cosine waves can be considered as sine waves shifted by $\frac{\pi}{2}$.

As can be seen from Figure 1.1, the sum of cosine waves closely approximates the ideal periodic square pulse as n increases.

In general, a periodic function $p(x)$ with period T_0 can be represented by a Fourier series:

$$
p(x) = \frac{a_0}{2} + \sum_{k=1}^{+\infty} [a_k \cos k\omega_0 x + b_k \sin k\omega_0 x] \quad (1.2)
$$

where ω_0 is the fundamental frequency equal to $\frac{2\pi}{T_0}$. The amplitudes of the sinusoids are given by the integrals:

$$
a_k = \frac{2}{T_0} \int_{-T_0/2}^{T_0/2} p(x) \cos(k\omega_0 x) dx \quad k = 0, 1, 2, \ldots \quad (1.3)
$$

$$
b_k = \frac{2}{T_0} \int_{-T_0/2}^{T_0/2} p(x) \sin(k\omega_0 x) dx \quad k = 0, 1, 2, \ldots \quad (1.4)
$$

Equation 1.2 can be more concisely represented by using the complex numbers properties, summarized in the following box:

Properties of Complex Numbers

$j = \sqrt{-1} = e^{j\pi/2}$

$e^{j\theta} = \cos\theta + j\sin\theta$

$e^{j(a+b)} = e^{ja}e^{jb}$

$\cos\theta = \frac{e^{j\theta}+e^{-j\theta}}{2}$

$\sin\theta = \frac{e^{j\theta}-e^{-j\theta}}{2j}$

If $z = Ae^{j\theta}$, then $|z| = A$, $arg(z) = \theta$

If $z = x + jy$, then $|z| = \sqrt{x^2 + y^2}$, $arg(z) = \tan^{-1}\left(\frac{y}{x}\right)$

$z^* = x - jy = Ae^{-j\theta}$ (complex conjugate of z)

In this case, Equation 1.2 becomes:

$$p(x) = \sum_{k=-\infty}^{+\infty} s_k e^{j\omega_0 kx} \tag{1.5}$$

where

$$c_k = \frac{1}{T_0} \int_{-T_0/2}^{T_0/2} p(x)e^{-j\omega_0 kx}\,dx. \tag{1.6}$$

This representation is often used in describing MRI because a complex exponential conveniently represents the precession of the magnetization vector.

The above expression for approximating a periodic function with a sum of sinusoids, assumes the following hypotheses (called Dirichlet's conditions) for the periodic function:

1. It has a finite number of discontinuities in the period T_0.

2. It and its derivative are piecewise continuous in the period T_0.

3. Its modulus has finite integral in the period T_0, that is, $\int_{-T_0/2}^{T_0/2} |p(x)|dx < \infty$.

In MRI, as in many other fields, signals or images can be assumed to be periodic or quasi periodic (if a signal, or an image, is not periodic, it has been demonstrated it can be decomposed to become periodic [88],[122]), at least for the purposes of image formation.

For completeness we also describe how to represent a nonperiodic function.

A nonperiodic function $f(x)$ can be represented as an integral of sinusoids of proper frequencies, amplitudes, and phase:

$$f(x) = \frac{1}{2\pi} \int_{-\infty}^{\infty} F(\omega) e^{j\omega x} d\omega \tag{1.7}$$

described by $F(\omega)$, the spectrum of the function $f(x)$ (the spectrum is represented by using the capital letter). Note that the sum has been replaced by an integral, that is, a continuous function of frequencies rather than just multiples of one fundamental.

The spectrum $F(\omega)$ can be computed from a signal using the Fourier transform:

$$F(\omega) = \int_{-\infty}^{\infty} f(x) e^{-j\omega x} dx \tag{1.8}$$

The derivation of Equations 1.7 and 1.8 is extensively treated in [13] and [12].

You will find many variations of Equations 1.7 and 1.8: for example, the normalizing factor, $\frac{1}{2\pi}$, can be placed before Equation 1.8 or before Equation 1.7; also, one can choose to work in normal frequency rather than angular frequency, then one can drop the $\frac{1}{2\pi}$ all together. Finally, one can reverse the signs of the exponents in the two transforms.

To illustrate the mathematics of the Fourier transform (the Fourier transform operator is abbreviated as FT), and to compare it with the previous periodic example, let us calculate the Fourier transform of a nonperiodic square pulse:

$$\prod(x) = \begin{cases} A & |x| < \frac{1}{2} \\ \frac{A}{2} & |x| = \frac{1}{2} \\ 0 & |x| > \frac{1}{2} \end{cases} \tag{1.9}$$

(shown in Figure 1.2). The FT of $\prod(x)$ can be calculated as follows:

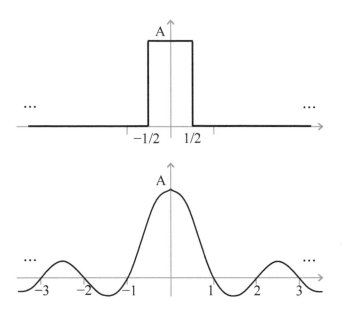

Figure 1.2
A non periodic square pulse (top) and the *sinc* function (bottom): they form
a Fourier pair, that is, one is the FT of the other.

$$
\begin{aligned}
\int_{-\infty}^{\infty} A \prod (x)\, e^{-j\omega x}\, dx
&= A \int_{-\frac{1}{2}}^{\frac{1}{2}} e^{-j\omega x}\, dx \\[2mm]
&= A \int_{-\frac{1}{2}}^{\frac{1}{2}} \left[\cos \omega x - j \sin \omega x\right] dx \\[2mm]
&= A \left. \frac{\sin \omega x}{\omega} \right|_{-\frac{1}{2}}^{\frac{1}{2}} \\[2mm]
&= A \left[\frac{\sin\left(\frac{1}{2}\omega\right)}{\omega} + \frac{\sin\left(\frac{1}{2}\omega\right)}{\omega} \right] \\[2mm]
&= A \frac{\sin\left(\frac{1}{2}\omega\right)}{\frac{1}{2}\omega} \\[2mm]
&= A \operatorname{sinc} \nu
\end{aligned}
\tag{1.10}
$$

where we have made use of the identity:

$$
\operatorname{sinc} x = \frac{\sin \pi x}{\pi x}.
\tag{1.11}
$$

The sinc function equals zero for all integer values of x, except $x = 0$, where it is assumed its value is 1, that is, the value of its limit for $x \to 0$. The amplitude of the oscillation decreases as x moves from the origin and it tends to 0 at ∞.

A plot of the sinc function is shown in Figure 1.2 (bottom).

The sinc is thus the FT of the nonperiodic square pulse and vice versa; they constitute an FT pair.

The FT of some other basic functions warrants consideration.

We start with the impulse function $\delta(x)$. It has the property to be zero everywhere except at the origin:

$$\delta(x) \;=\; 0 \qquad x \neq 0. \tag{1.12}$$

Moreover, its integral equals to 1:

$$\int_{-\infty}^{\infty} \delta(x)\,dx = 1. \tag{1.13}$$

We can imagine the δ function to be the limit of a sequence of nonperiodic square pulses of unit area where the width of the base of the pulses becomes narrower and moves toward zero but, at the same time, its amplitude increases to maintain the area constant and equal to 1.

According to the previous equations, we can define

$$\int_{-\infty}^{\infty} \delta(x - x_0)\,p(x)\,dx = p(x_0) \tag{1.14}$$

where $p(x)$ is an arbitrary function, continuous at x_0. Equation 1.14 helps in illustrating the δ as a function whose amplitude equals its integral (its shape has no physical meaning: only its integral counts), with the effect that integrated with a suitable function $p(x)$, the result is the value of $p(x)$ at $x = x_0$; the sifting property of the δ function [12].

Application of Equation 1.14 straightforwardly yields the FT of many important functions.

If we consider the function

$$h(x) = s\delta(x), \tag{1.15}$$

its FT can be easily derived using Equation 1.14:

$$H(\omega) = \int_{-\infty}^{\infty} s\delta(x)\,e^{-j\omega x}\,dx = se^0 = s. \tag{1.16}$$

Thus, the Fourier decomposition of a single spike at the origin consists of equally weighted sinusoids at all frequencies, represented in the FT of $\delta(x)$ by a constant amplitude s at all ω.

The FT^{-1} of $H(\omega)$ is given by:

$$h(x) = \int_{-\infty}^{\infty} s e^{j\omega x}\,d\omega = \int_{-\infty}^{\infty} s\cos(\omega x)\,d\omega + j\int_{-\infty}^{\infty} s\sin(\omega x)\,d\omega. \tag{1.17}$$

Because the second integrand is an odd function, the integral is zero; the first integral can be approached using distribution theory [12], with the result that:

$$h(x) = s \int_{-\infty}^{\infty} \cos(\omega x) d\omega = s\delta(x). \tag{1.18}$$

This last result asserts that the FT of a constant function is a δ at the origin: a constant function does not vary in time or space, and hence, does not contain sinusoids with non-zero frequencies. The amplitude of the δ, s, has a clear physical meaning; it represents the area of the analyzed curve.

The sine function $h(x) = A\sin(\omega_0 x)$ is a continuous periodic function to which FT can be applied as follows:

$$
\begin{aligned}
H(\omega) &= \int_{-\infty}^{\infty} A\sin(\omega_0 x)e^{-j\omega x} dx \\
&= \frac{A}{2} \int_{-\infty}^{\infty} \left[e^{j\omega_0 x} - e^{-j\omega_0 x} \right] e^{-j\omega x} dx \\
&= \frac{A}{2}j \int_{-\infty}^{\infty} \left[-e^{-jx(\omega-\omega_0)} + e^{-jx(\omega+\omega_0)} \right] dx \\
&= \frac{A}{2}j \left[\delta(\omega + \omega_0) - \delta(\omega - \omega_0) \right],
\end{aligned}
\tag{1.19}
$$

obtained by using the same considerations used to calculate Equation 1.18.

In an analogous way, if $h(x) = A\cos(\omega_0 x)$, its FT is the following:

$$
\begin{aligned}
H(\omega) &= \int_{-\infty}^{\infty} A\cos(\omega_0 x)e^{-j\omega x} dx \\
&= \frac{A}{2} \int_{-\infty}^{\infty} \left[e^{j\omega_0 x} + e^{-j\omega_0 x} \right] e^{-j\omega x} dx \\
&= \frac{A}{2}j \int_{-\infty}^{\infty} \left[e^{-jx(\omega-\omega_0)} + e^{-jx(\omega+\omega_0)} \right] dx \\
&= \frac{A}{2}j \left[\delta(\omega - \omega_0) + \delta(\omega + \omega_0) \right],
\end{aligned}
\tag{1.20}
$$

both the FT of $\sin \omega_0 x$ and of $\cos \omega x$ are composed two δ functions, one at $-\omega_0$ and the other at $+\omega_0$. This is intuitive because the FT of a function is an expansion of it in terms sinusoids: expanding either a single sine or a single cosine in terms of sines and cosines yields the original sine or cosine function.

Note, however, that the FT of a sine consists of one negative and one positive δ, whereas the FT of a cosine results into two positive δ. This is because the sine is an odd function $(\sin(-\omega x) = -\sin(\omega x))$, whereas the cosine is an even function $(\cos(-\omega x) = \cos(\omega x))$.

Comparing the formula for the FT in Equation 1.8 with the formula for the inverse FT in Equation 1.7 (in the following indicated by FT^{-1}), we observe that they differ only in the sign of the argument to the exponential: FT and FT^{-1} are qualitatively the same.

Thus, if we know the transform from the function domain to the frequency domain, we also know the transform from the frequency domain to the function domain.

Another useful function is the Gaussian function $s(x) = e^{-\pi x^2}$. The FT of a Gaussian is a Gaussian. As an example, this result is demonstrated in the following box:

Fourier Transform of a Gaussian Function

$$FT\left[s(x)\right] = \int_{-\infty}^{\infty} e^{-\pi x^2} e^{-j2\pi x\nu} dx$$

$$= \int_{-\infty}^{\infty} e^{-\pi(x^2 + j2x\nu)} dx$$

$$= e^{-\pi\nu^2} \int_{-\infty}^{\infty} e^{-\pi(x+j\nu)^2} dx$$

by placing $\sqrt{\pi}(x + j\nu) = y$, we have

$$= e^{-\pi\nu^2} \frac{1}{\sqrt{\pi}} \int_{-\infty}^{\infty} e^{-y^2} dy$$

$$= e^{-\pi\nu^2}$$

Note that $\int_{-\infty}^{\infty} e^{-y^2} dy = \sqrt{\pi}$.

The last analyzed function is that represented by a series of equispaced δ functions, which is sometimes referred to as the *shah* function (also called Dirac comb). It is defined as follows:

$$shah(x) = \sum_{n=-\infty}^{\infty} \delta(x - n\Delta x), \tag{1.21}$$

where Δx represents the distance between consecutive δ.

It can be demonstrated that the FT of a *shah* function is a *shah* function [12].

This function is particularly useful, in the following section, when discussing sampling.

The shape of the functions analysed above are shown in Figure 1.3 (a more extensive gallery can be found in [12]).

The spectrum of a function tells what frequencies are present in the function and their relative amount. Rapid changes imply high frequencies; gradual changes imply low frequencies. The zero frequency component, or dc term, is the average value of the function.

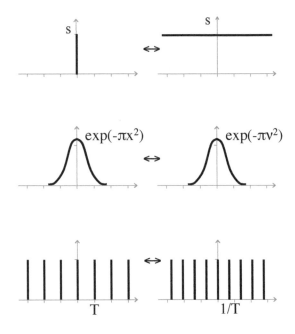

Figure 1.3
Some Fourier pairs. From top to bottom: $\delta(x)$, e^{-x^2} and $shah(x)$.

An important concept is that of a *band-limited function*. A function is band-limited if its spectrum has no frequencies above some maximum frequency, that is, its spectrum occupies a finite interval of frequencies, not the entire frequency range.

To summarize: the key point of this section is that a function can be easily converted from its domain (that is, a function of x) to the frequency domain (that is, a function of ν or ω), and vice versa. Thus, a function can be interpreted in either of two domains: the function or the frequency domain.

The Fourier transform and the inverse Fourier transform can be used to switch between the two domains.

1.2   FT Properties

Here we illustrate the most important properties of the FT operator.

In most real-world applications, including MRI, the generic variable x used above is better substituted with t to indicate that the function originates as a signal evolving in time. The function and Fourier domains then are those of time and frequency, respectively. A second Fourier pair arises in image analysis between the spatial and spatial frequency domains; in this case, the variable x is used to indicate space.

A useful aspect of FT pairs is that mathematical operations performed in one domain have corresponding operations in the other; as previously reported, to switch from one space to the other to make easier a certain operation before to return to the original space. Some of the presented properties are demonstrated; others are only reported.

For clarity, lowercase and uppercase functions are FT pairs, for example $S(\nu) = FT\left(s(t)\right)$.

1. **Linearity.** Taken two constants, a and b, and two functions, $s(t)$ and $z(t)$, linear combinations are maintained from one space to the other, that is:

$$as(t) + bz(t) \Longleftrightarrow aS(\nu) + bZ(\nu)$$

2. **Conjugation and Reflection.**

   (a) Complex conjugation (the symbol * is used for the conjugation operator):

   $$s^*(t) \Longleftrightarrow S^*(-\nu)$$

   (b) Reflection:

   $$s(-t) \Longleftrightarrow S(-\nu)$$

   (c) Hermitian conjugation:

   $$s^*(-t) \Longleftrightarrow S^*(\nu)$$

   (d) Symmetries:
   * If $s(t)$ is purely real, then $S(\nu)$ is hermitian:

   $$s(t) = s^*(t) \Longleftrightarrow S(\nu) = S^*(-\nu)$$
   * If $s(t)$ is purely imaginary, then $S(\nu)$ is antihermitian:

   $$s(t) = -s^*(t) \Longleftrightarrow S(\nu) = -S^*(-\nu)$$
   * If $s(t)$ is even, then $S(\nu)$ is even:

   $$s(t) = s(-t) \Longleftrightarrow S(\nu) = S(-\nu)$$
   * If $s(t)$ is odd, then $S(\nu)$ is odd:

$$s(t) = -s(-t) \iff S(\nu) = -S(-\nu)$$

* If $s(t)$ is hermitian, then $S(\nu)$ is purely real:

$$s(t) = s^*(-t) \iff S(\nu) = S^*(\nu)$$

* If $s(t)$ is antihermitian, then $S(\nu)$ is purely imaginary:

$$s(t) = -s^*(-t) \iff S(\nu) = -S^*(\nu)$$

3. **Scaling.** Given $a \in \mathbb{R}$ and $a \neq 0$, then

$$s(at) \iff \frac{1}{|a|}S(\tfrac{\nu}{a})$$

It can be easily shown:

$$
\begin{aligned}
FT\left[s\left(at\right)\right] &= \int_{-\infty}^{\infty} s\left(at\right) e^{-j2\pi\nu t}\,dt \\
&= \frac{1}{|a|}\int_{-\infty}^{\infty} s\left(k\right) e^{-j2\pi\frac{\nu}{a}k}\,dk \\
&= \frac{1}{|a|}S\left(\frac{\nu}{a}\right).
\end{aligned}
$$

From the *time scaling property*, it is evident that if the width of a function is decreased while its height is kept constant, then the FT becomes wider and shorter. If its width is increased, its transform becomes narrower and taller. A similar *frequency scaling property* is given by

$$\frac{1}{|a|}s\left(\tfrac{t}{a}\right) \iff S\left(a\nu\right).$$

4. **Shifting.**[1]

   (a) Time shift ($t_0 \in \mathbb{R}$):

   $$s(t - t_0) \iff e^{-j2\pi\nu t_0} S(\nu)$$

   (b) Frequency shift ($\nu_0 \in \mathbb{R}$):

   $$e^{j2\pi\nu_0 t}s(t) \iff S(\nu - \nu_0)$$

---

[1]Note the opposite signs in the exponentials in the two domains.

We can obtain the result for the time shift as follows:

$$
\begin{aligned}
FT\left[s\left(t-t_0\right)\right] &= \int_{-\infty}^{\infty} s\left(t-t_0\right) e^{-j2\pi\nu t}\,dt \\
&= \int_{-\infty}^{\infty} s\left(k\right) e^{-j2\pi\nu(k+t_0)}\,dk \\
&= e^{-j2\pi\nu t_0} \int_{-\infty}^{\infty} s\left(k\right) e^{-j2\pi\nu k}\,dk \\
&= e^{-j2\pi\nu t_0} S(\nu).
\end{aligned}
$$

The *time shifting* property means that the FT of a shifted function is just the transform of the unshifted function multiplied by an exponential factor having a linear phase.

Analogously for the *frequency shifting* property, but in this case the exponential multiplicative term, in the time domain, has the opposite sign.

This property is particularly useful for MRI to represent phase shift due to residual magnetic field inhomogeneity (see Chapter 3 and Chapter 4).

5. **Differentiation.**[2]

    (a) Time differentiation:

$$
\tfrac{d}{dt}s(t) \Longleftrightarrow j2\pi\nu S(\nu)
$$

    (b) Power scaling or frequency differentiation:

$$
ts(t) \Longleftrightarrow \tfrac{-1}{j2\pi}\tfrac{d}{d\nu}S(\nu)
$$

6. **Convolution.** Consider two functions $s(t)$ and $h(t)$, we say that a third function $g(t)$ is the convolution between $s(t)$ and $h(t)$, indicated by $g(t) = s(t) \otimes h(t)$, iff

$$
g(t) = \int_{-\infty}^{\infty} s(k)h(t-k)dk \tag{1.22}
$$

For the convolution function, the following property applies:

$$FT\left[s(t) \otimes h(t)\right] = S(\nu)H(\nu).$$

In fact,

---

[2]Note the opposite signs in the two domains.

$$
\begin{aligned}
G(\nu) &= FT\left[s(t) \otimes h(t)\right] \\
&= FT\left[\int_{-\infty}^{\infty} s\left(k\right) h\left(t-k\right) dk\right] \\
&= \int_{-\infty}^{\infty}\left[\int_{-\infty}^{\infty} s\left(k\right) h\left(t-k\right) dk\right] e^{-j2\pi\nu t} dt \\
&= \int_{-\infty}^{\infty} s\left(k\right)\left[\int_{-\infty}^{\infty} h\left(t-k\right) e^{-j2\pi\nu t} dt\right] dk \\
&= H(\nu)\int_{-\infty}^{\infty} s\left(k\right) e^{-j2\pi\nu k} dk \\
&= H(\nu)S(\nu).
\end{aligned}
$$

Using a similar derivation, it can be shown that the FT of a product is given by the convolution of the individual transforms, that is, $FT\left[s(t)h(t)\right] = S(\nu) \otimes H(\nu)$.

This property is very useful as it allows us to calculate the convolution between two functions easily: in the dual space, this corresponds to a simple product. In fact, to calculate the convolution between two time domain functions, we have to proceed by calculating the FT of the two functions, then the product of these in the frequency domain and, finally, the FT^{-1}.

For example, what is the spectrum of the function resulting from convolving two nonperiodic square pulses? Convolving two functions corresponds to multiplying their spectra, therefore, convolving two nonperiodic square pulses corresponds to the multiplication of two sinc functions. Similarly, the convolution of n square pulses corresponds to the *sinc* raised the the nth power. If n goes to infinity, the convolution of n nonperiodic square pulses approaches a Gaussian. An example of an FT pair obtained by convolution is that reported in Figure 1.4 for the nonperiodic triangle function.

This operation is also known as *filtering.*

A *low-pass filter* attenuates high frequencies, ideally leaving all frequencies below a *cut-off* frequency and removes all frequencies above it. Thus, in the frequency domain, an ideal low-pass filter is a nonperiodic square pulse.

A *high-pass filter* attenuates low frequencies, ideally completely removing all frequencies below the cut-off frequency, having a shape of a constant minus a nonperiodic square pulse.

A *band-pass filter* preserves a range of frequencies relative to those

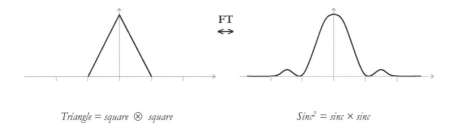

Triangle = square ⊗ square   *Sinc² = sinc × sinc*

Figure 1.4
The nonperiodic triangle function, can be obtained through convolution of two nonperiodic squares, in the signal domain, or through the multiplication of two sinc functions, in the Fourier domain. The Fourier pair has not been obtained as a direct FT of the original functions.

outside that range removing all frequencies outside its band and corresponds to a paired square pulse.

Removing high frequencies leaves the more slowly varying components of the signal and results in a blurry function. Removing low frequencies, on the other hand, enhances the high frequencies and creates a sharper function containing rapidly changing edges.

7. **Correlation.** The correlation integral, like the convolution integral, has important theoretical and practical applications. The correlation integral is defined as

$$h(t) = \int_{-\infty}^{\infty} s(k)g(t+k)dk \qquad (1.23)$$

and like the convolution function, it forms a FT pair given by

$$FT\left[h(t)\right] = S(\nu)G^*(\nu).$$

This is the statement of the *correlation theorem*. If $s(t)$ and $g(t)$ are the same function, the above integral is normally called *autocorrelation* function, and *cross correlation* if they differ [13]. The FT pair for the autocorrelation is simply

$$FT\left[\int_{-\infty}^{\infty} s(k)s(t+k)dk\right] = |S(\nu)|^2.$$

8. **Parseval's identity.** Parseval's identity, also noted as *Parseval's theorem*, states that the power of a signal represented by a function

$h(t)$ is the same whether computed in signal space or in frequency space, that is:

$$\int_{-\infty}^{\infty} h^2(t)dt = \int_{-\infty}^{\infty} |H(\nu)|^2 d\nu.$$

The *power spectrum* of the function $H(\nu)$, indicated with $PW\,[H(\nu)]$, is given by

$$PW\,[H(\nu)] = |H(\nu)|^2, \quad -\infty \leq \nu \leq +\infty.$$

The power spectrum is used in MRI to define the information content of a projection (Chapter 5).

In summary, a function can be decomposed into a sum (Fourier series) or an integral (FT) of harmonics.

The set of the frequencies of the harmonics, contained in the original function, along with their relative amplitudes and phases constitute the frequency content of the function.

There are functions for which the FT does not exist; however, most physical functions have a FT, especially if the transform represents a physical quantity.

In the next section, the sampling conditions will be discussed; it is very important for digital representation of functions and, in particular, of MRI signals and images.

1.3   Sampling, Interpolation, and Aliasing

The frequency space and convolution are useful to describe the processes of sampling and interpolation. Here we use the term "interpolation" to indicate the process to reconstruct the original continuous function from the sampled one. We do not use the term "reconstruction" to avoid confusion with the image reconstruction techniques discussed in the next chapters.

In the time domain, sampling a continuous function $h(t)$ can be viewed simply as multiplying it by the *shah* function defined in Equation 1.21. Since the *shah* function is zero everywhere except on integer values, the multiplication retains just the information at the sampling points.

On these points, the sampled function has the same values of the original function. Sampling in the time domain is illustrated in Figure 1.5, up direction; a Gaussian function, $h(t)$, has been multiplied by a *shah* function to produce a point-wise Gaussian function.

The sampling operation can be also performed in the frequency domain,

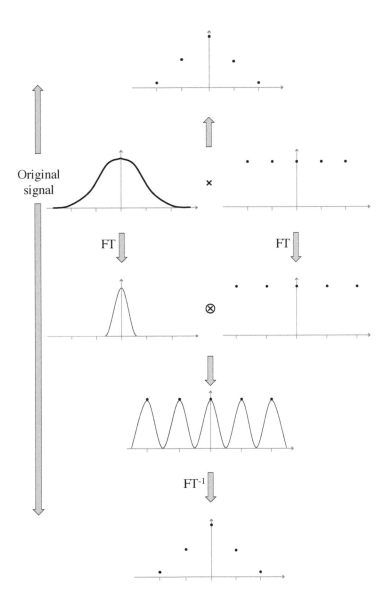

Figure 1.5
Sampling of a function: in the time domain (up direction), the function is multiplied by a *shah* function; in the frequency domain (down direction), sampling is obtained through convolution of the FT functions, and the result is transformed back in the time domain.

but recalling the convolution property, multiplication in the time domain corresponds to convolution in the frequency domain. As the FT of a *shah* function is a *shah* and the FT of a Gaussian function is a Gaussian function, the convolution causes the spectrum $H(\omega)$ of the original (Gaussian) function to be replicated (each copy of the spectrum is centred on a δ). The sampling procedure in the frequency domain is illustrated in Figure 1.5, down direction.

Though sampling can be performed in both domains, it is more efficient (computationally) if applied in the time domain.

Now, let us consider the interpolation process to recover the original continuous function from the sampled one.

The interpolation process can also be performed in both spaces, as illustrated in Figure 1.6. In contrast to the sampling process, interpolation is more efficient (computationally) if performed using the frequency domain.

In the frequency domain, the sampling process resulted in the replication of the spectrum of the original function. For nonoverlapping replicas, as is the case illustrated in Figure 1.6, the original function can be recovered through the application of an ideal low-pass filter (this role is assumed by the nonperiodic square pulse), thus removing all extra copies of the spectrum $H(\omega)$. Less efficiently, we can perform the same operation in the time domain. In this case, however, we must convolve the samples with a *sinc* function (the FT^{-1} of the nonperiodic square pulse). Thus, the *sinc* function has an interpolation role that serves to recover the continuous function from a set of its samples. This is illustrated in Figure 1.6 up direction, for a Gaussian function.

The result of sampling/interpolation is the original function. That is, no information was lost with sampling. This result is known as the *sampling theorem*: a signal can be reconstructed from its samples without loss of information, if the original signal has no frequencies above $\frac{1}{2}$ the sampling frequency. This limit is also called Nyquist's limit or the *Nyquist frequency*. Formally, If $2B_\nu$ is the function bandwidth and ν_s is the sampling frequency, then $2B_\nu < \nu_s$ must hold to avoid aliasing.

To prove that, consider two continuous signals: a continuous signal $f(t)$ and the $shah(t)$ function. Let the result of the multiplication be

$$f_s(t) = f(t)shah(t) = f(t) \sum_{n=-\infty}^{\infty} \delta(t - n\Delta t)$$

where Δt is the distance between two consecutive samples. By taking the Fourier transform and applying the multiplication/convolution property,

$$F_s(\omega) = FT\left(f_s(t)\right) = FT\left(f(t)\right) \otimes FT\left(shah(t)\right) = \int_{-\infty}^{\infty} F_s(\tau)shah(\tau - \omega)\, d\tau.$$

By using the shifting property of the function $shah(t)$, the integral can be removed, as in

$$F_s(\omega) = \int_{-\infty}^{\infty} F_s(\tau)shah(\tau - \omega)\, d\tau = F_s(\tau)shah(\tau - \omega)$$

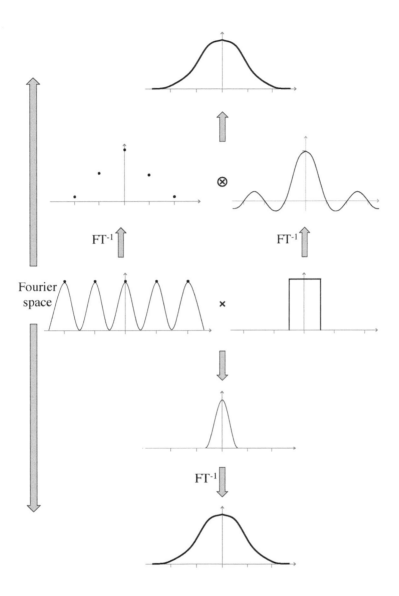

Figure 1.6
The interpolation process to recover a continuous function from its samples: in the time domain (up direction), the sampled function is convolved with the *sinc* function; in the Fourier domain (down direction), the FT of the sampled function is multiplied by a nonperiodic square function, and the result is transformed back in the time domain.

and, finally, by expanding the definition of $shah(t)$,

$$F_s(\omega) = F(\omega) \sum_{n=-\infty}^{\infty} \delta(\omega - n\nu_s) \tag{1.24}$$

where $\nu_s = \frac{1}{\Delta t}$ is the sampling frequency. The final result is a summation of shifted $F(\omega)$. Let $\nu_{\max}$ be the maximum frequency of $F(\omega)$, then $F(\omega)$ is bounded on $[-\nu_{\max}, \nu_{\max}]$. The bandwidth of $F(\omega)$ is then $2\nu_{\max} = 2B_\nu$. To avoid overlap for a replicated $F(\omega)$ shifted by ν_s, the condition $\nu_{\max} < \frac{1}{2}\nu_s$, that is, $2B_\nu < \nu_s$, must hold.

To derive the *sampling theorem*, a condition must hold: the original continuous (analog) function (signal) has to be band-limited. A sampled band-limited continuous (analog) function (signal) can be perfectly reconstructed from an infinite sequence of samples if the sampling frequency exceeds $2B_\nu$ samples per second, where $B_\nu = \nu_{\max}$ is the highest frequency in the original signal.

If a signal contains a component at exactly B_ν hertz or higher, then samples spaced at exactly $\frac{1}{2B_\nu}$ seconds do not completely determine the signal.

But, what happens if the condition of the *sampling theorem* is not satisfied? The sampling/interpolation process can fail!

When a function is sampled at less than its Nyquist frequency, the spectrum replicas may overlap. When overlap occurs, it is impossible to recover the original function because the contribution of some frequencies will be summed to other frequencies: this effect is called *aliasing*. This will occur also for any function that is not band-limited.

The result is that the reconstruction process is unable to differentiate between the original spectrum and the aliased spectrum and, hence, the low-pass filter process yields an aliased spectrum that misrepresents some frequencies with the effect that the function cannot be perfectly reconstructed. This effect is sketched in Figure 1.7.

As aliasing is an important concept that recurs in this book, it is best to illustrate it with an example also in the signal domain.

Consider a sinusoid with a frequency of 10 Hz (corresponding to 10 cycles per second). Now sample it at 9 Hz. The sampling frequency is lower than the frequency of the function and aliasing occurs. This is reported in Figure 1.8.

The figure shows that sampling $\sin(2\pi 10t)$ at 9 Hz is the same as sampling the function $\sin(2\pi t)$ at 9 Hz: aliasing has been produced at the frequency of 1 Hz, which is the difference between the signal frequency and the sampling frequency.

The *sampling theorem* leads to a formula for the reconstruction of the original signal.

However, it makes an idealization in that it only applies to functions that are sampled for infinite time and, at the same time, are band limited (they

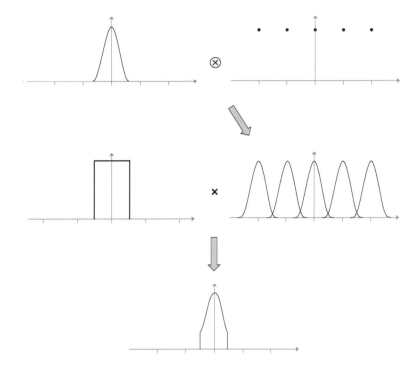

Figure 1.7

Aliasing due to undersampling: resulting Fourier spectrum replicas overlap in some frequencies and the resulting filtered spectrum differs from the original.

have a finite maximum frequency). As explained by Fourier theory, temporally limited functions cannot be perfectly band-limited.

Perfect interpolation is mathematically possible for the idealized model, but it is impossible for real-world signals: for these last signals, interpolation allows only an approximation, albeit in practice often a very good one.

The constructive proof of the theorem allows also us to understand how aliasing can occur when sampling does not satisfy its hypotheses.

It is important to note, however, that the sampling theorem provides a sufficient condition, not a necessary one, for perfect interpolation (reconstruction). The field of compressed sensing, as discussed in Chapter 5, provides a looser sampling condition applicable when the underlying signal is known to be sparse and sampling is random, not regular. Knowledge of the sparsity (or compressibility) of the signal and random sampling can be used to provide lower limits on the number of samples needed to allow perfect signal recon-

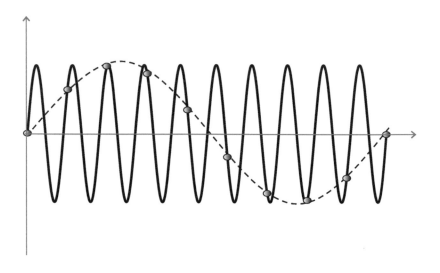

Figure 1.8
The original function $\sin(2\pi 10t)$ (continuous line), a frequency of 10 Hz, is subsampled at 9 Hz. This is like sampling the function $\sin(2\pi t)$, a frequency of 1 Hz, at 9 Hz. The frequency of the original signal is greater than the sampling frequency, thus aliasing is generated at 1 Hz, the frequency difference between the signal frequency and the sampling frequency. The resulting sampled signal is a "ghost" at 1 Hz (the dotted line is used to highlight the resulting wave).

struction. In particular, compressed sensing accepts a sub-Nyquist sampling criterion. However, signal reconstruction in compressed sensing needs non-linear solvers. As will be further discussed in Chapter 5, besides compresses sensing, other sub-Nyquist sampling criteria can be efficiently used.

1.4   Instruments for Image Analysis

In image analysis, methods to compare reconstructed or processed images are necessary. To this end, we can divide methods into those that rely on subjective criteria or those that rely on objective criteria [37].

Frequently, the images would undergo human visual interpretation and analysis (human opinion). For this reason, a subjective image evaluation is common and can be very appropriate. A method that is currently used to compare images processed with different strategies is visual comparison performed by a series of experts. The experts evaluate a series of characteristics by choosing, for each of them, among fixed adjectives (very low, low, poor, sufficient, good, very good); as a summary metric, the mean value of these subjective evaluations is typically used.

This approach can be very useful because it ensures relevant information, that is, the details to which experts give weight, is considered. Thus, small details can be more significant than extended features of the same image.

Visual comparison is used at points in this book for differently processed images but in general is left to the reader (i.e., the images are always reported – at least the most significant). Although the summary metric may be expressed as a number, it does not follow a proper additive scale (good + good $\neq$ very good). Moreover, it cannot be considered objective.

When the quantification is expressed in terms of a mathematical function of the image values themselves, we say that it corresponds to an objective criterion. Here, we define the objective measures that are most used in this book.

Many measurements are based on the power spectrum (indicated as the square of the FT) of noise ($n(x, y)$) and of the nondegraded image ($f(x, y)$). One of these is the *signal-to-noise ratio* (SNR), approximated in the frequency domain by using the following equation:

$$SNR = \frac{1}{M * N} \frac{\sum\limits_{u=1}^{M} \sum\limits_{v=1}^{N} |F(u,v)|^2}{\sum\limits_{u=1}^{M-1} \sum\limits_{v=1}^{j=N} |No(u,v)|^2} \tag{1.25}$$

where $F(u,v)$ and $No(u,v)$ are the FT of the original image and of the noise, respectively, and M and N are the number of rows and columns (to avoid confusion with the number of columns, we use No to define the FT of noise). Images with low noise content will have high SNR; conversely, images with high noise will have low SNR.

The previous definition is very useful if we consider both the original image and the noise as known functions (this can occur when we use artificial noise added to theoretical signals or images).

In other cases, when we compare a processed image $\widehat{f}$ with the theoretical image f, we can consider a statistical measurement of SNR calculated directly in the signal/image domain, as follows:

$$SNR = \frac{\sum\limits_{x=1}^{M} \sum\limits_{y=1}^{N} \widehat{f}(x,y)^2}{\sum\limits_{x=1}^{M} \sum\limits_{y=1}^{N} \left[f(x,y) - \widehat{f}(x,y)\right]^2}. \tag{1.26}$$

where M and N are defined as above. This definition considers the difference $\widehat{f}(x,y) - f(x,y)$ as a noise; in this case, the ratio is between an image, $\widehat{f}(x,y)$, and a noise, $\widehat{f}(x,y) - f(x,y)$.

Throughout the book we use one or the other definition of SNR, depending on the convenience: if we have to add a noise to the original image (or to a signal), we use the former; if want to evaluate the result of an image (or a signal) processing procedure, we use the latter.

When compression methods are used, as for some of the reconstruction methods used in Chapter 5, *peak signal-to-noise ratio* (PSNR) is commonly used as a measure of the image quality after lossy compression (to calculate the behavior of a lossy compression algorithm). To define PSNR, it is necessary to calculate the *mean square error* (MSE):

$$MSE = \frac{1}{M * N} \sum_{x=1}^{M} \sum_{y=1}^{N} \left[f(x,y) - \widehat{f}(x,y) \right]^2. \tag{1.27}$$

PSNR is calculated as follows:

$$PSNR = \frac{max(\widehat{f})^2}{MSE}. \tag{1.28}$$

Different reconstruction methods can be compared on the same image: PSNR is not an absolute measure. Greater is PSNR value, better the reconstruction will resemble the original image.

It is worth noting that both SNR and PSNR can also be expressed in decibels (dB), as follows:

$$SNR = 10 log(SNR) \tag{1.29}$$

and

$$PSNR = 10 log(PSNR) = 10 log \left(\frac{max(f)^2}{MSE} \right) = 20 log \left(\frac{max(f)}{\sqrt{MSE}} \right) \tag{1.30}$$

where the base of the logarithm is 10. The measurement in decibels is useful when images (or signals) have very high dynamic range. An increment of 0.25 dB is usually considered the minimum improvement detectable by the human eye. We will use the absolute measurement or the dB measurement depending on convenience; when the dB measurement is used, the suffix "dB" is added; elsewhere, a pure number is reported.

2

MRI: Conventional Imaging Techniques and Instruments

CONTENTS

Magnetic resonance imaging has radically modified the practice of medicine in general and radiology in particular. It is a computer-based imaging modality that displays the body in thin tomographic slices, based on the interaction between radio waves and nuclei composing the object being scanned (whether an inanimate sample or a living subject, often called *sample*), in the presence of a static magnetic field. Physical characteristics of a volume element or *voxel* of tissue are translated by the computer into a two-dimensional image composed of picture elements, or *pixels*. Pixel intensity in MRI reflects spin density, generally as the hydrogen of water, weighted by some parameters such as chemical shift, relaxation times, magnetic susceptibility, motion, etc. MRI can produce images of very good morphological quality and/or very high functional interest. This reflects not only the contrast between different tissues, due to physicochemical properties of the tissues, but also the ability to display changes in contrast over time. Moreover, MRI allows images to be collected directly in any plane, that is, the usual axial, sagittal, coronal, or any degree of obliquity. This chapter contains some basic concepts regarding MR signal formation and conventional imaging techniques. In this setting, the

MR signal arises in the time domain, and the image occupies the frequency domain.

The role of the static magnetic field, magnetic field gradients and radio frequency pulses, imaging principles, bandwidth, and sampling are discussed. Magnetic field gradients, in particular, are analyzed in terms of slice selection, frequency encoding, and phase encoding. Finally, both the design of an experimental equipment and of a MRI numerical simulator are described.

This chapter is intended as background for understanding the innovative encoding techniques presented in the following chapters. The discussion is restricted to the essential phenomenological and macroscopic concepts. For our purposes it does not constitute a limitation. Detailed discussion of such themes can be found in References [48], and [10], and in Reference [140].

2.1   Magnetic Resonance Phenomenon

Much of the mathematical description of magnetic resonance can be found in the pioneering works of Purcell [112], Bloch [11], and Hahn [42]. Rather than repeat the detailed mathematical relationships here a more conceptual working description will be given.

Each proton plus neutron contained in the nucleus of an atom has a magnetic dipole moment, and in turn the nucleus as a whole has a net magnetic moment $\boldsymbol{\mu}$ that depends on the number of protons and neutrons it contains.

A second property of the constituent protons plus neutrons that is reflected in the nucleus as a whole is a form of angular momentum called *spin* (for a description of the nature of spin, see "The story of the spin" [135]). The total angular momentum of a nucleus, commonly called the *nuclear spin* ($\boldsymbol{I}$), is also related to the number of constituent protons plus neutrons.

$\boldsymbol{I}$ can have values of 0, half-integers, or whole integers. The quantized value of $\boldsymbol{I}$ determines the interaction of a nucleus with the magnetic fields it encounters. For nuclei with even atomic weight and even atomic number, $\boldsymbol{I} = 0$. Such nuclei are unaffected by a magnetic field and cannot be observed by magnetic resonance. Nuclei having either integer or half-integer spin numbers are affected by magnetic fields and can be detected by magnetic resonance techniques. Nuclei of interest for purposes of magnetic resonance imaging have half-integer spin numbers. ^{1}H, ^{17}O, ^{19}F, ^{23}Na, and ^{31}P are some nuclei of interest in medicine and biology. The hydrogen nucleus ^{1}H, in particular, consisting of a single proton, due to its large magnetic moment and its high natural abundance in tissues that are largely composed of water, is most often the nucleus of choice in MR imaging and spectroscopy. Hydrogen nuclei are also referred as; "nuclei," "water," and "spins". These terms can all be used interchangeably.

For nuclei with spin $\boldsymbol{I} = 1/2$, the interaction of the nuclear moment with an external magnetic magnetic field is very similar to that of a bar magnet, composed of a north pole and a south pole. In the presence of the earth's magnetic field, the bar magnet's north pole points toward the earth's south pole, while its south pole points toward the earth's north pole. In the same way, a nuclear magnetic moment $\boldsymbol{\mu}$ in a magnetic field $\mathbf{B}_0$ points along the direction of the applied magnetic field. The magnetic moment $\boldsymbol{\mu}$ is a vector quantity because it has a magnitude and a direction.

For normal biological tissues, the hydrogen nuclei are randomly oriented when the object to be scanned is not located in a magnetic field (Figure 2.1(a)).

In this situation, the net magnetic moment for the protons in the tissue as a whole is zero.

When the object to be scanned is placed in an external magnetic field $\mathbf{B}_0$, measured in Tesla (T) or in Gauss (1T= 10000 Gauss; for comparison, the earth's magnetic field intensity is about 0.5 Gauss), the nuclei align with the external field, either in the direction of the magnetic field ("parallel"), or in the direction opposite to the magnetic field ("antiparallel") (Figure 2.1(b)). The parallel orientation corresponds to a lower energy state: for this reason, more nuclei are oriented in this direction than in the opposite direction. The sum of all of the magnetic moment vectors of the individual nuclei then yields a nonzero net magnetization vector $\mathbf{M}$. Over time, $\mathbf{M}(t)$ represents changes in the average orientation and distribution of the nuclear magnetic moments in the object to be scanned.

The spin and magnetic moment have a particular relationship. In the absence of a magnetic field, both are randomly distributed among the population of spins in the sample. In this condition, the net magnetic moment formed by the vector sum of the $\boldsymbol{\mu}$ of individual nuclei is zero ($\mathbf{M} = 0$).

In the presence of a magnetic field, however, the spins are forced away from an orientation perpendicular to the magnetic field, causing them to have individual z components (if z is the direction of the magnetic field) either aligned with (parallel) or against (antiparallel) the magnetic field, with a low prevalence through parallel direction due to energy. The magnetic moments of the nuclei thus distributed yield a nonzero net $\mathbf{M}$.

Further, the spins experience a torque that causes them to precess (and so, the magnetic moments), as shown in Figure 2.2. Under equilibrium conditions, however, the transverse component distribution of the magnetic moments is uniform and the precession is not observable at the bulk scale.

When the equilibrium is disturbed, the temporal evolution of the net magnetization vector reflects the cumulative behavior of the individual moment vectors in terms of orientation and distribution.

The angular frequency of the precession, ω_0, depends on the strength of the applied magnetic field $\mathbf{B}_0$, according to the Larmor equation:

$$\omega_0 = \gamma B_0 \qquad (2.1)$$

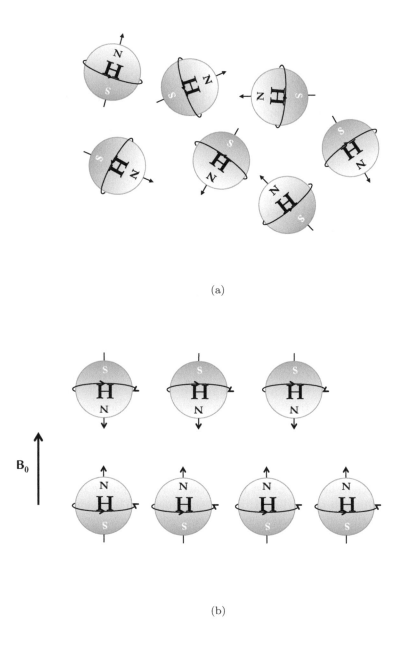

Figure 2.1
Spin orientation in absence (a) and in presence (b) of an external magnetic field.

Precessional motion

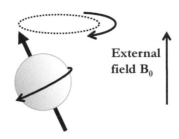

External field B_0

Figure 2.2
Precessional motion of a spin subjected to the external magnetic field.

where γ is a constant, the gyromagnetic ratio, whose value is characteristic of the nucleus of interest (for hydrogen, $\gamma = 267.53 \cdot 10^6$ rad s^{-1} T^{-1}).
It is important to note that Equation 2.1 can be written in terms of linear frequency, by expressing γ in Hertz T^{-1} (for Hydrogen, $\gamma = 42.60 \cdot 10^6$ Hz T^{-1}).

2.1.1   References Frames

The orientation of the static magnetic field, $\mathbf{B_0}$, is conventionally taken to be along the z-axis of a three-dimensional Cartesian coordinate system (Figure 2.3).

The z-axis, also called the "longitudinal" direction, is shown in the up position. The plane perpendicular to the z-axis forms the "transverse" xy plane.

The motion of the net magnetization vector $\mathbf{M}$, in response to a perturbing secondary magnetic field, is difficult to follow in the static Cartesian coordinates system, also called *laboratory frame*, that provides the viewpoint of an observer in a laboratory where both are stationary (Figure 2.4(a)).

Relative to an observer, the nuclei subjected to a magnetic field along the z-axis precess around the z-axis at the Larmor frequency (see Figure 2.4(a)). If a secondary magnetic field perturbs the nuclei, changing their orientation relative to the z-axis, their precession then describes complex spiral patterns.

To simplify the task of describing the motion of the magnetic moments, one can use a rotating reference frame (see Figure 2.4(b)). In this description, the x' and y' axes rotate around the z axis at the Larmor frequency of the

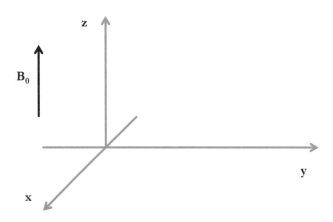

Figure 2.3
Static magnetic field direction with respect to the reference coordinate system.

nuclei in the external static $\mathbf{B}_0$. To an observer, "riding" on the x' axis, the spins appear stationary, and the laboratory appears rotating.

The rotating reference frame eliminates the spin at the Larmor frequency so that one can describe small changes in the precession frequency of nuclei without having to keep up with the Larmor frequency. The spiral trajectory that accompanies the perturbation by a secondary field, for instance, becomes a single rotation (α).

The use of the rotating frame and of the net magnetization vector $\mathbf{M}$ provides a convenient way of describing what happens to the magnetization with time.

2.1.2 Excitation and Resonance

When a sample exposed to a static magnetic field $\mathbf{B}_0$ arrives at the equilibrium, $\mathbf{M}$ is aligned with the z-axis. If an external magnetic field perturbs the nuclei, $\mathbf{M}$ loses its alignment with the z-axis.

This is accomplished by applying a radio frequency (RF) pulse. The RF pulse produces a magnetic field perpendicular to the z-axis, $\mathbf{B}_1$, which causes the nuclei to change their orientation with respect to their original orientation.

The change of orientation is particularly efficient when the frequency of the RF pulse matches the Larmor frequency exactly (i.e., the RF pulse has to be in resonance with the Larmor frequency).

If the RF pulse is at a different frequency, resonance will not occur, with

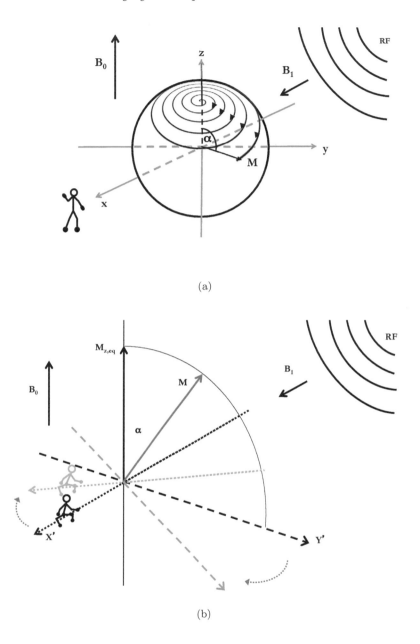

(a)

(b)

Figure 2.4
Magnetization evolution during an RF pulse: (a) in the laboratory frame; (b) in the rotating frame. In the laboratory frame, an observer sees the magnetization vector to precess around the field direction. In the rotating frame, the observer rotates along with the magnetization vector, at the same frequency, and he sees just the tilting movement of the magnetization vector. For simplicity, the rotating field $\mathbf{B}_1$ was represented as fixed.

the result that there is no effect on the net magnetic moment and it does not move from its stable state along the z axis.

If one looks at what occurs from the standpoint of the rotating frame, it is evident that if the $\mathbf{B}_1$ is applied at the Larmor frequency along the x'-axis, around which the spins will display a precession, it appears as a stationary magnetic field (see Figure 2.4(b)). Thus, meanwhile the $\mathbf{B}_1$ field is applied, the net magnetic moment rotates relative to the z-axis.

The angle α, through which the sample magnetization is rotated by the RF pulse, is proportional both to the $\mathbf{B}_1$ amplitude and pulse duration τ, $\alpha \propto B_1\tau$.

Any angle can be obtained through a proper combination of timing and intensity of the RF pulse at the Larmor frequency. This angle is usually called the "tip," "flip," or "tilt" angle and is important in the design of pulse sequences.

Useful values of α for imaging purposes are the angles $90°$ $(\pi/2)$, $180°$ (π), and $270°$ $(3\pi/2)$.

By convention, the net magnetic vector is considered in terms of its component vectors, $\mathbf{M}_z$ and $\mathbf{M}_{xy}$ (see Figure 2.5). As the vector $\mathbf{M}$ moves away from alignment in the z direction, the $\mathbf{M}_z$, or longitudinal component decreases while the $\mathbf{M}_{xy}$, or transverse component, increases. When an RF pulse flips $\mathbf{M}$ into the xy plane, the $\mathbf{M}_z$ component vanishes and the $\mathbf{M}_{xy}$ component is maximized and equal to the net magnetization vector $\mathbf{M}$.

The amplitude of the magnetic field $\mathbf{B}_1$, produced by the RF pulse, is much smaller than the static field. The resonant effect of $\mathbf{B}_1$ allows $\mathbf{B}_1$ of a few Gauss to disturb the alignment of the magnetization $\mathbf{M}$ even though the static magnetic field $\mathbf{B}_0$ is thousands of Gauss.

When the RF pulse is switched off, the magnetization vector gradually returns to its equilibrium orientation. Relaxation, discussed in the following subsection, shows the reduction of the $\mathbf{M}_{xy}$ component and the restoration of $\mathbf{M} = \mathbf{M}_z$ during decay. This is such a magnetic field oscillating at the Larmor frequency ω_0 in the static laboratory frame (Figure 2.6), and with no oscillation in the rotating frame (Figure 2.7) [49, 50].

This oscillation in laboratory frame can be observed by induction of current in a neighboring conductor coil following Lenz's law. This signal is called a *free induction decay* (FID) when the RF pulse is turned off (Figure 2.6 and Figure 2.7). The reduction of $\mathbf{M}_{xy}$, which was 0 at equilibrium before the RF pulse, leads to the produced signal which is a decaying signal.

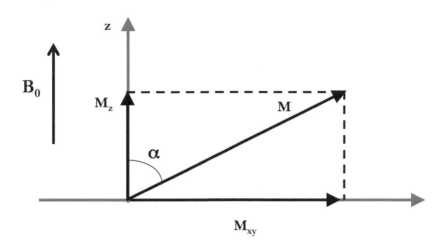

Figure 2.5
Magnetization vector position tilted of an angle α with respect to the z axis. The vector components are also reported.

2.1.3 Relaxation of Magnetization

At equilibrium, $\mathbf{M}_z$, the component of the magnetic vector in the z direction, is at a maximum, and $\mathbf{M}_{xy} = 0$ because the nuclei precess about the z-axis at the Larmor frequency with no phase coherence (see Figure 2.8).

The return at equilibrium is referred to as relaxation and can be considered in terms of $\mathbf{M}_z$ and $\mathbf{M}_{xy}$ changes, processes described as longitudinal and transversal relaxation, respectively.

The physicochemical environment can influence the longitudinal and transversal relaxation properties. The impact of relaxation of $\mathbf{M}$ is reflected in the signals acquired in MRI.

The temporal behavior of $\mathbf{M}$ can be effectively described through the solution of phenomenological equations that were introduced by Felix Bloch in 1946 [11] and, in his honor, called Bloch equations. The Bloch equations are used to calculate the nuclear magnetization $\mathbf{M}$ as a function of time in presence of relaxations (here we use their solutions to show longitudinal and transversal relaxations; in the last section of this chapter, their solution are used to implement an MRI numerical simulator).

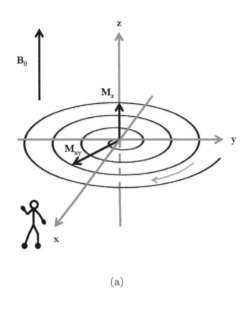

(a)

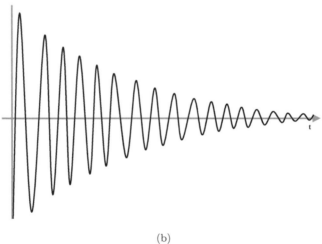

(b)

Figure 2.6
Transversal magnetization rotation in the laboratory frame: (a) it describes a spiral shape; (b) the decaying signal oscillating at the Larmor frequency generated by the spiral movement.

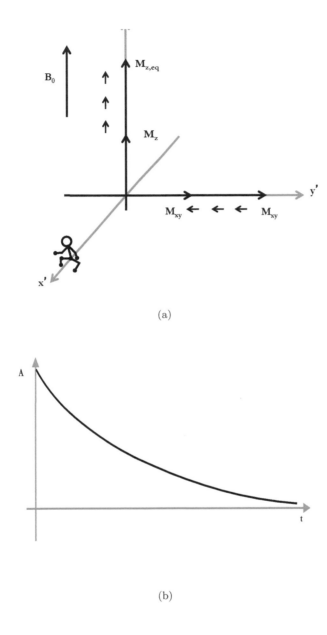

(a)

(b)

Figure 2.7
Transversal magnetization in the rotating frame. (a) magnetization components change their lengths with time: the transversal component reduces,
while longitudinal component increases. (b) the decaying signal, produced
by transversal component. As can be seen, it does not oscillate.

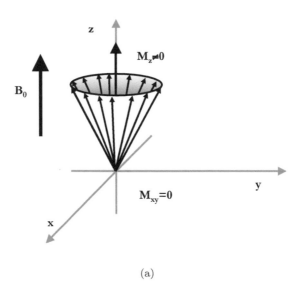

(a)

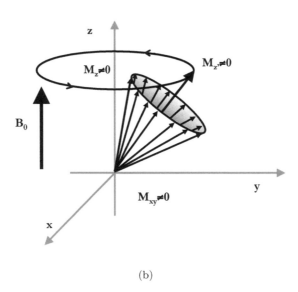

(b)

Figure 2.8
Magnetization vector at the equilibrium (a), where the transverse component is zero and the component along the magnetic field direction is maximum, and after an RF pulse. After the RF pulse (b), both the magnetization components are different from zero.

2.1.3.1   Longitudinal Relaxation

Immediately after the application of an RF pulse, (say of 90°), the net magnetization vector lies in the xy plane, that is, $\mathbf{M}_{xy} \neq 0$ and $\mathbf{M}_z = 0$ ($\mathbf{M}_{xy}$ and $\mathbf{M}_z$ can be both present, depending on the value of the flip angle α, as shown in Figure 2.5).

After excitation, as the system relaxes to its equilibrium state, the $\mathbf{M}_{xy}$ component decays to 0 and the $\mathbf{M}_z$ component gradually recovers the equilibrium state along the z direction (Figure 2.7(a)).

Viewed from the rotating frame (Figure 2.7(a)), after a flip angle α, the $\mathbf{M}_z$ component evolves as:

$$\mathbf{M}_z(t) = \mathbf{M}_{z,eq} - [\mathbf{M}_{z,eq} - \mathbf{M}_z(\alpha)]e^{-t/T_1}. \tag{2.2}$$

where $\mathbf{M}_z(\alpha)$ represents the residual $\mathbf{M}_z$ component after an RF pulse of angle α has been turned off and t is time ($t = 0$ corresponds to the moment of switching off the RF pulse). T_1 is the decay constant for the recovery of the z component of the nuclear spin magnetization, $\mathbf{M}_z$, toward its thermal equilibrium value, $\mathbf{M}_{z,eq}$. This phenomenon is called *longitudinal relaxation*, and T_1 is called *longitudinal relaxation time* (see Figure 2.9, for a 90° RF pulse).

Specific cases of notice include

1. If $\alpha = 90°$, $\mathbf{M}$ has been tilted into the xy plane, $\mathbf{M}_z(0) = 0$ and the recovery is simply

$$\mathbf{M}_z(t) = \mathbf{M}_{z,eq}\left(1 - e^{-t/T_1}\right). \tag{2.3}$$

   Such experiments (*saturation recovery*) form the basis of a discriminating technique: *spin echo* imaging.

2. If $\alpha = 180°$, the initial magnetization is inverted, $\mathbf{M}_z(0) = -\mathbf{M}_{z,eq}$, and the recovery is

$$\mathbf{M}_z(t) = \mathbf{M}_{z,eq}\left(1 - 2e^{-t/T_1}\right). \tag{2.4}$$

   Such experiments, where $\alpha = 180°$ (*inversion recovery*), are commonly used to measure T_1 values.

The value of T_1 reflects the interaction of nuclei with their molecular environment, and specifically how quickly the energy can be transferred from the nuclei to the environment, or "lattice" (for this reason T_1 is also called *spin − lattice relaxation time*). Longitudinal relaxation involves redistributing the populations of the nuclear spin states to reach the thermal equilibrium. By definition, this is not energy conserving.

T_1 can be used to characterize materials and changes in the biological

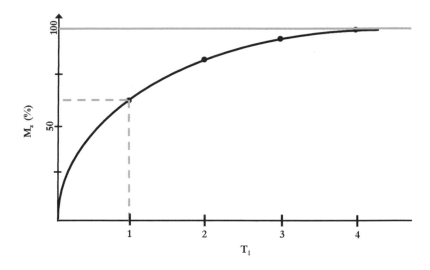

Figure 2.9
Longitudinal magnetization recovery after a 90° RF pulse. The recovery is reported as a function of T_1.

environment within an organ caused by various pathologies can be measured through changes in T_1 values.

The presence of other substances in a sample greatly reduced T_1, that is, longitudinal relaxation is speeded up.
Although indicated as a constant in the previous equations, T_1 also depends on the strength of the magnetic field in which the object being scanned is placed.

2.1.3.2 Transverse Relaxation

At equilibrium, in the presence of a static magnetic field, $\mathbf{B}_0$, the nuclear precession about the z axis is incoherent, that is, the nuclei are not in phase (Figure 2.8).

A rapid RF pulse at the Larmor frequency, producing a flip angle α, causes the nuclei to precess in phase (Figure 2.8(b)). Over time this phase coherence is gradually lost, an effect called *transverse relaxation*, whose time constant T_2 describes phase dispersion, the loss of phase coherence in the xy plane. The equation describing the evolution of the transverse magnetization component is

$$\mathbf{M}_{xy}(t) = \mathbf{M}_{xy}(\alpha)e^{-t/T_2} \tag{2.5}$$

where $\mathbf{M}_{xy}(t)$ is the transverse magnetization at the time t and $\mathbf{M}_{xy}(\alpha)$ is

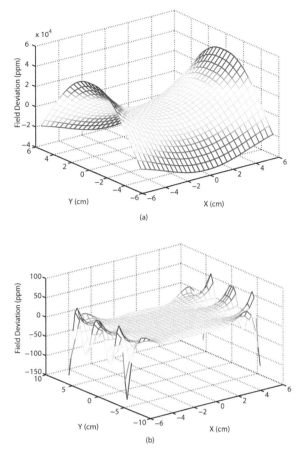

(a)

(b)

FIGURE 3.3
Residual magnetic field inhomogeneity for a dedicated permanent magnet: (a) before shimming, (b) after shimming.

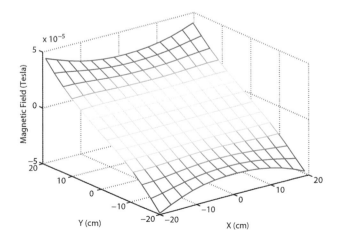

FIGURE 3.5
An experimental field gradient shape whose maximum inhomogeneity, 4%, is localized to the corners.

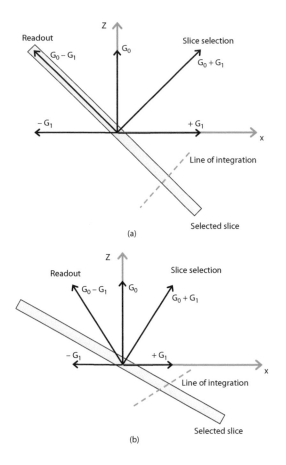

FIGURE 4.2
Slant slice imaging [32] due to residual inhomogeneity (a) with adjustable gradients of equal strength to the permanent gradient and (b) with adjustable gradients of smaller strength than the permanent gradient. In these figures, $\mathbf{G}_y$ is considered to be orthogonal to the xz-plane. Line of integration has also been indicated.

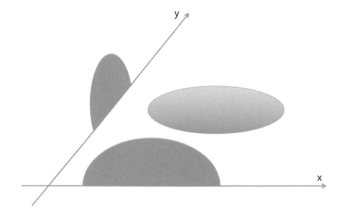

FIGURE 5.1
Internal sample symmetries, known in advance, allow exact reconstruction from just few projections (in this case, one projection would suffice).

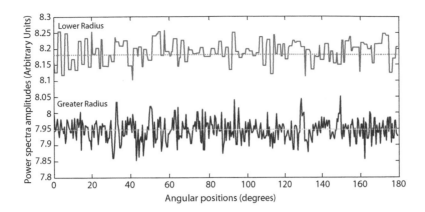

FIGURE 5.10
Power spectrum plot of the two circular paths outlined in Figure 5.9(b). The corresponding mean values are represented by dotted lines.

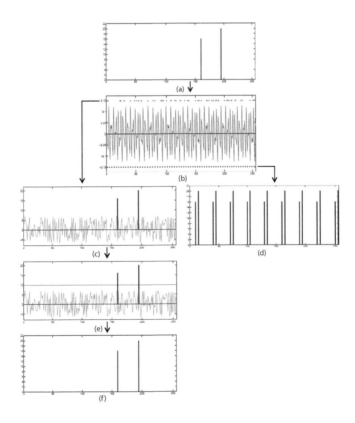

FIGURE 5.13
A signal (a) is sparsely undersampled (b) both in random way (upper part dots) and in regular way (lower part dots). Random undersampling aliasing (c) is in the form of random noise and can be eliminated through thresholding (e) to obtain a correct signal recovery (f). Regular undersampling aliasing (d) is in the form of unavoidable repetitions. Similar to Figure 5 in [74], different data were used.

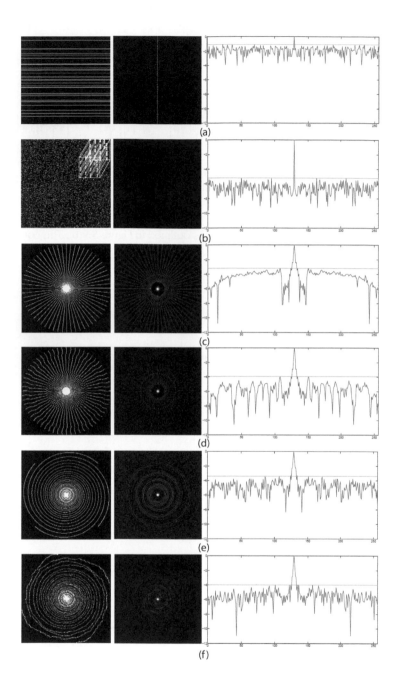

FIGURE 5.14
Sampling trajectories (left): (a) random parallel lines, (b) random points in a cross section of random lines in 3-D, (c) uniform radial lines, (d) perturbed uniform radial lines, (e) variable density spirals, and (f) variable density per- turbed spirals. PSFs (center) and PSFs most significant directions (right). Horizontal line indicates the coherence level. Similar to Figure 6 in [74], dif- ferent data were used.

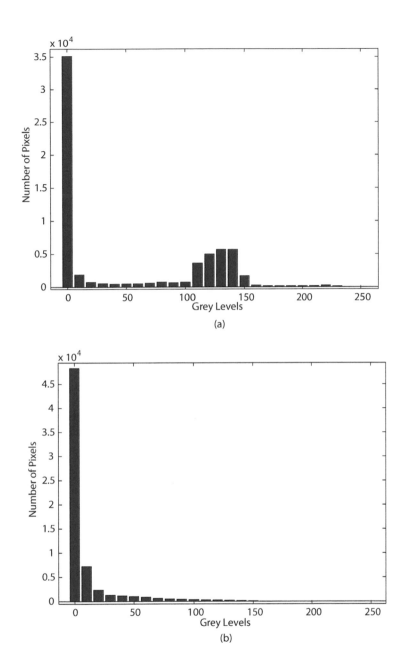

FIGURE 5.16
Histograms of (a) the image in Figure 5.15(a) and (b) of its gradient transform Figure 5.15(b).

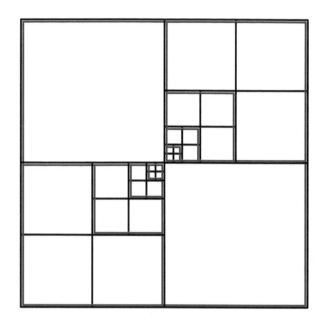

FIGURE 5.18
Subwindows adaptive acquisition scheme: an example.

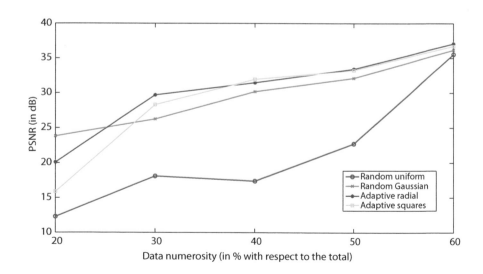

FIGURE 5.19
PSNR values of the CS reconstructions by different acquisition modalities. The data set numerosity ranged from 20% of the total to 60% of the total in steps of 10%.

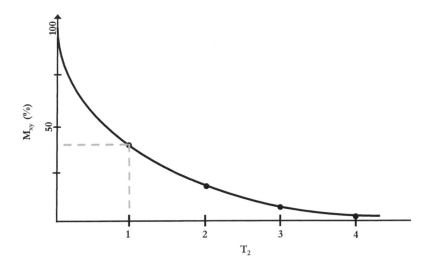

Figure 2.10
Transverse magnetization recovery after a 90° RF pulse. The recovery is reported as a function of T_2.

the initial transverse magnetization at $t = 0$, the time in which the RF pulse was turned off (Figure 2.10, for a 90° RF pulse). As the phase coherence of the nuclear spins present at $t = 0$ is lost, $\mathbf{M}_{xy}$ tends to vanish.

Transverse relaxation corresponds to a growing decoherence of the transverse nuclear spin magnetization that relates closely to the behavior of the local magnetic field. Two potential contributions to T_2 relaxation are the magnetic field oscillations at the Larmor frequency due to neighboring precessing spins (*spin–spin relaxation*) and the random variations in the local magnetic field, which causes the rate of precession to vary between different spins.

For this reason, transverse relaxation is also called spin–spin relaxation. T_2 is usually much less dependent on field strength, $\mathbf{B}_0$, than T_1.

The sensitivity of T_2 to perturbing magnetic fields extends to static magnetic field inhomogeneity. In this case, the effects are described by a time constant known as T_2^*, which has important implications in imaging, and the observed rate of transverse relaxation may be much shorter than the theoretical value.

Loss of coherence due to static magnetic field inhomogeneity is not, however, a true "relaxation" process; it is not random, but depends on the location

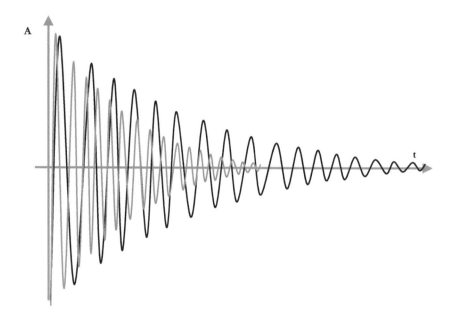

Figure 2.11
Comparison of FID signals duration in absence of magnetic field inhomogeneities (black signal) and in presence of magnetic field inhomogeneities (gray
signal).

of the molecule in the magnet. The relationship between T_2 and T_2^* is

$$\frac{1}{T_2^*} = \frac{1}{T_2} + \frac{1}{T_{inhom}} = \frac{1}{T_2} + \gamma \Delta B_0 \qquad (2.6)$$

where γ is the gyromagnetic ratio and ΔB_0 is magnetic field variation.

In a real system, T_2^* is always shorter than T_2 (the FID duration is reduced,
as shown in Figure 2.11), and T_2 is generally shorter than T_1, but this is not
always true. The relationship that always holds is $T_2 \leq sT_1$ for $s = 2$.

As an exercise, we demonstrate why s cannot be greater than 2. We start
by remarking that longitudinal and transversal relaxations are two distinct
aspects of the same phenomenon, that is, each one is regulated by a proper
relaxation time, but both are subject to the rule that the amplitude of the
magnetization vector they produce cannot be greater than the amplitude of
the magnetization vector at the equilibrium, $\mathbf{M}_{z,eq}$.

In fact, after an RF pulse (assuming, without loss of generality, a 90° RF pulse for simplicity), Equations 2.2 and 2.5 hold. By calculating the modulus of the resulting vector and substituting $T_2 = sT_1$, we get

$$|\mathbf{M}(t)| = M_{z,eq} \left[e^{-2t/(sT_1)} + \left(1 - e^{-t/T_1} \right)^2 \right]^{1/2} =$$

$$= M_{z,eq} \left[e^{-2t/(sT_1)} + 1 - 2e^{-t/T_1} + e^{-2t/(T_1)} \right]^{1/2} \leq M_{z,eq}.$$

What is included into square brackets has to be lower than or equal to 1 to ensure that $|\mathbf{M}(t)| \leq M_{z,eq}$ for all t. If this last condition does not hold, we could have that, for some time instant t_p, longitudinal component is equal to $\mathbf{M}_{z,eq}$ and a residual, nonzero, transversal component $\mathbf{M}_{xy}(t_p)$ is still present. The fact that the longitudinal magnetization has recovered the equilibrium value implies the energy balance is recovered: the transverse component has to be zero. For this reason, the following must holds:

$$e^{-2t/(sT_1)} + 1 - 2e^{-t/T_1} + e^{-2t/(T_1)} \leq 1. \tag{2.7}$$

By separating some terms and multiplying by $e^{t/(T_1)}$, we get

$$e^{-2t/(sT_1)+t/T_1} + e^{-2t/(T_1)+t/T_1} \leq 2$$

that is,

$$e^{(-t/T_1)(-1+2/s)} + e^{-t/T_1} \leq 2. \tag{2.8}$$

Equation 2.8 is true for all $t \geq 0$ if $2/s - 1 \geq 0$, that is, for $s \leq 2$, and so $2T_1 \geq T_2$.

The magnetization evolution, in a phenomenological sense, due to both transverse and longitudinal relaxations is depicted in Figure 2.12 for the case of $T_1 > T_2$.

After RF excitation, see Figure 2.12 upper row, both transverse and longitudinal relaxations are active and in the second phase, see Figure 2.12 second row, just residual longitudinal relaxation is effective.

T_2 and T_1 are equal for pure water.

T_1, T_2, T_2^*, blood flow, proton density, etc., form the basis for the contrast available in MRI; the richness of parameters makes MRI a very special imaging modality.

2.1.4 Signal Detection

As described above, when a 90° RF pulse is switched off, the transverse magnetization will decay to the original 0 value by emitting a FID. The FID is not spatially differentiated: its amplitude is proportional to the cumulative spin population in the whole sample.

A 90° RF pulse causes the net magnetic vector to "flip" into the transverse

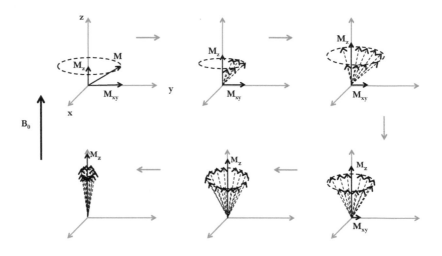

Figure 2.12
Relaxations dynamics for $T_1 > T_2$. In the upper row, after RF excitation, both transverse and longitudinal relaxations are active. In the latter row, only residual longitudinal relaxation remains to more slowly restore its equilibrium value.

plane. Considering the behavior of the net magnetization from the viewpoint of the laboratory frame, one realizes that, at the moment the 90° pulse ends, the magnetization vector is left rotating about the z-axis in the xy plane at the Larmor frequency. If a coil is placed so that its axis lies in the transverse plane (i.e., in the xy plane), then an electromotive force (of the order of millivolts), oscillating at the Larmor frequency, is induced on the coil (Figure 2.13)[1].

---

[1]In our discussion, the assumption is that the material being observed contains only hydrogen in completely identical chemical and magnetic environments, at a single Larmor frequency. All the nuclei experience the same magnetic field and therefore produce a FID having a single frequency. However, if some of these identical nuclei are to experience different magnetic fields, either due to some interaction within the sample or because a spatially varying magnetic field (magnetic field gradient) has been applied to the sample, the signal induced in the receiver coil will be more complex. The reason for this is that nuclei at different locations experience different magnetic field values, which in turn causes these nuclei to precess at different Larmor frequencies. In this circumstance, the signal detected by the receiving coil is the superposition of the individual Larmor frequencies from nuclei at different locations.

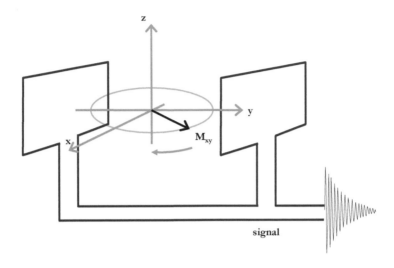

Figure 2.13
A receiving coil is placed perpendicularly to the longitudinal magnetization plane, thus ensuring an oscillating signal be induced in the coil by transverse magnetization precession.

This coil is essentially a radio antenna. The induced signal is proportional to the number of excited nuclei in the object being scanned, and persists only as long as there is a transverse component of the net magnetization vector (i.e., detection occurs when $\mathbf{M}_{xy} \neq 0$). Transverse magnetization results when an RF pulse tips the longitudinal magnetization away from the z-axis toward the transverse xy plane. A 90° RF pulse tips the magnetization all the way into the xy plane; a 180° RF pulse (twice as strong or twice as long as a 90° pulse) tips the magnetization, so it is pointing down, along the z-axis.

A 90° pulse converts all longitudinal magnetization to transverse magnetization, as a 180° pulse does not generate transverse magnetization, yielding the maximum transverse magnetization and thus the maximum signal. Flip angles lower than 90° do not cause a complete orientation of the magnetization, and therefore they produce less transverse magnetization, that is, less signal. However, since less time is needed for longitudinal recovery, they can be repeated rapidly, and generate more signals per time unit, see Figure 2.14. This is the basis for rapid imaging.

Besides FID, another type of MR signal can be produced: an echo. When a 90° pulse and a 180° pulse are applied sequentially, a spin-echo signal is generated (Figure 2.15). After a 90° pulse, transverse magnetization decays

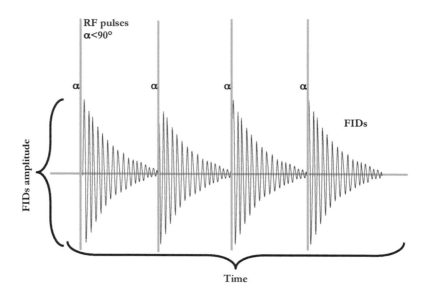

Figure 2.14
Generation of a series of signals caused by a train of consecutive RF pulses.
α is the the magnetization vector inclination with respect to the z axis, after
an RF pulse.

rapidly with T_2^* as protons get out of phase (i.e., "lose phase coherence") due
to inhomogeneity in the main magnetic field.

The effect of the 180° pulse is to "refocus" the phase of the protons, causing
them to regain coherence and thereby to recover transverse magnetization,
thus producing the spin echo. Following the spin echo, coherence is again lost
as the protons continue to precess at slightly different frequencies.

If another 180° pulse is applied, coherence can be reestablished in a second
spin echo.

Multiple spin echo signals can be produced if the original 90° pulse is
followed by multiple 180° pulses. This "echo train" is illustrated in Figure 2.16.

Although the 180° pulses cause rephasing of phase coherence lost due to
fixed inhomogeneities in the main field, complete rephasing is not possible
due to randomly fluctuating magnetic fields within the object being scanned
itself and the loss of transverse magnetization to longitudinal recovery. Thus,
the maximum intensity of the spin echo signals in the echo train is limited
by an exponentially decaying curve (Figure 2.16). The time constant of this
decay curve is the second magnetic relaxation time T_2. T_2^* is always shorter

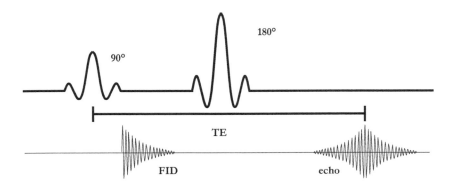

Figure 2.15
Spin-echo sequence timing.

than T_2 because the former includes inhomogeneity in the magnet as well as randomly fluctuating internal fields in the substance. T_2 decay is only due to the fluctuating internal fields in the substance.

A conventional spin echo signal results from a 90°–180° RF pulse pair. Adjusting the time between the 90°–180° pulses dictates the location (the echo time TE) of the spin echo along the T_2 decay curve. Since the decay differs between tissues with different T_2 values, the TE modification allows us to establish T_2 weighting (contrast proportional to T_2).

Repeating the spin echo, 90°–180° pair has a similar weighty effect in regard to T_1 values of the tissues. When the time between successive 90° pulses is very long (repetition time, TR), all the excited spins will have returned to their equilibrium in $\mathbf{M}_z$. Those tissues with shorter T_1s, however, will effectively reach equilibrium before tissues with longer T_1s, and this for a TR chosen in this window of time yields greater signals.

An inversion recovery (IR) sequence results from a 180°–90° pulse pair. Since the final RF pulse in this IR sequence is a 90° pulse, a FID signal is produced. By adding a final 180° pulse, that is, 180°–90°–180°, an IR sequence can produce a spin echo signal.

Another way to produce an echo signal is through gradient inversion, thus producing a gradient-echo, but this is not been discussed here (detailed description can be found in [48, 10, 140]).

In the following section, we discuss the role of magnetic field gradients.

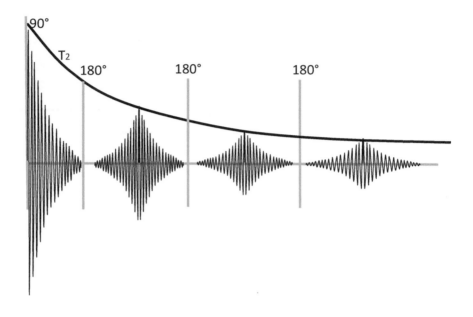

Figure 2.16
Generation of a series of echoes caused by a train of consecutive 180° RF
pulses, following a single 90° RF pulse. The amplitudes of the echoes dimin-
ishes with T_2.

2.2 Imaging Gradients

Magnetic field gradients play a special role in imaging. In a homogeneous,
static magnetic field, Equation 2.1 holds and as a consequence, there is no way
to differentiate contributions coming from different spatial districts (voxels).
Spatial discrimination is obtained through the use of magnetic field gradients.

A magnetic field gradient is a magnetic field which is directed along $\mathbf{B}_0$,
its amplitude is varied linearly with the position along a given axis, and it
sums with the main magnetic field $\mathbf{B}_0$ (Figure 2.17). Each imaging apparatus
is equipped with a set of three gradient coils, each generating a field gradient along one of the three orthogonal axes x, y, and z, respectively. Having
supposed $\mathbf{B}_0$ directed along the z axis, the three gradients are indicated as

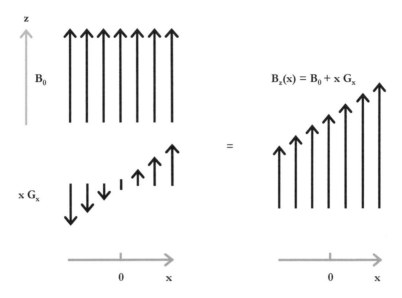

Figure 2.17
A magnetic field gradient is added to the main magnetic field: the resulting magnetic field is linearly variable with the position along the gradient direction.

follows:

$$
\begin{aligned}
\mathbf{G}_x &= \frac{\partial \mathbf{B}_0}{\partial x} \\
\mathbf{G}_y &= \frac{\partial \mathbf{B}_0}{\partial y} \\
\mathbf{G}_z &= \frac{\partial \mathbf{B}_0}{\partial z}
\end{aligned}
$$

$$(2.9)$$

The gradients can be combined to produce a magnetic field gradient along any desired spatial direction. For imaging purposes, the gradients are normally used in three different forms: frequency encoding, phase encoding, and slice selection. These are described in details.

2.2.1 Frequency Encoding

Frequency encoding is employed in almost all MRI experiments and is accomplished by applying a magnetic field gradient to the imaged object during signal acquisition to spatially modulate the frequencies in the signal as it is being readout. This gradient spatially encodes a signal by assigning a unique precession frequency to each spin position such that the induced time domain signal can be represented by the envelope of a superposition of frequencies, each corresponding to a different spatial location. The FT of the time domain signal reveals the frequency content, with each frequency being linearly related to a corresponding spatial location along the gradient direction. The amplitude at each frequency is mainly related to the corresponding spin density inside the spatial location affected by that frequency but may be modulated by other imaging parameters such as relaxation times, chemical shift, magnetic susceptibility, or the presence of paramagnetic substances.

Figure 2.18 shows the application of a $\mathbf{G}_x$ gradient along the x direction: the gradient causes the total magnetic field to change linearly along the x direction according to the relationship:

$$\mathbf{B}_z(x) = \mathbf{B}_0 + x \cdot \mathbf{G}_x. \tag{2.10}$$

As a consequence, also the resonance frequency becomes a linear function of the spatial position along x (gradient direction):

$$\omega(x) = \gamma(B_0 + x \cdot G_x). \tag{2.11}$$

As an example, consider three objects placed along the x axis in the positions x_1, x_2, and x_3, respectively, as illustrated in Figure 2.18.

In this case, the signal in presence of $\mathbf{G}_x$ is composed by three different frequencies corresponding to the three different spatial positions. In fact, the FT of the signal will produce a spectrum in which three peaks are present, each corresponding to a different sample. The frequency difference between peaks will depend on the physical separation between the objects and the magnetic field gradient value. The frequency encoding gradient can be applied along any physical direction.

2.2.2 Phase Encoding

The frequency gradient applied during readout recreates the spin distribution in one direction. A second spatial encoding can be obtained by applying a phase discriminating process. Phase encoding is so called because it relies on the phase of the spins being changed along the gradient direction. The process is described in Figure 2.19.

After an RF pulse, the spins of the sample can be considered to be in phase. At this point, if we turn on a magnetic field gradient, suppose directed

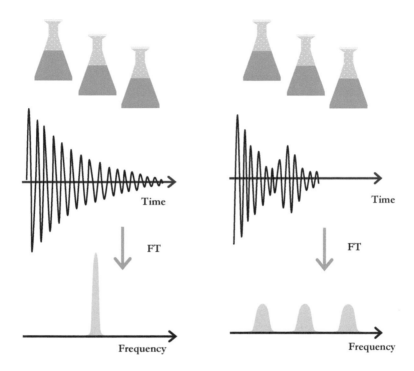

Figure 2.18
A sample, composed by three bottles is placed in a static magnetic field: with
no magnetic field gradient (left) the FT of the resulting signal is undifferenti-
ated; when a magnetic field gradient is added (right), an amplitude modulated
signal results which, after FT, reproduces three different peaks, one for each
bottle.

along y, the magnetic field difference along y will cause the spins to have
position-dependent precession frequencies. In a word, while the gradient is on,
the spins will dephase (in the same time, the slower spins will accumulate a
smaller phase angle than faster spins due to their lower frequency, i.e., angular
velocity). When the gradient is turned off, the original Larmor frequency is
restored (just $\mathbf{B}_0$ is present) but the spins will maintain the relative position-
dependent phase angle accumulated during the application of the gradient.

In a given position, having fixed the temporal duration of the gradient,
the accumulated phase dispersion depends on the $\mathbf{G}_y$ intensity:

$$\phi(y) = \int_{t_1}^{t_2} \gamma y G_y dt = \gamma y (t_2 - t_1) G_y. \tag{2.12}$$

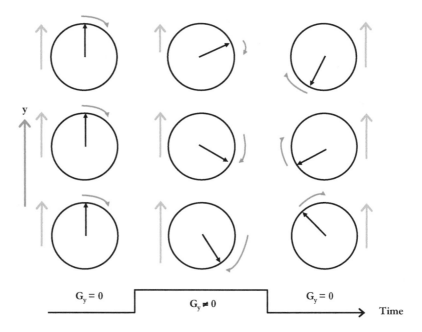

Figure 2.19
Phase-encoding gradient: when the gradient is off (left) the three voxels have
the same phase and frequency; during the gradient application (center), the
voxels accumulate a phase difference due to their different frequencies; after
the gradient is switched off (right), the voxels return to the same frequency,
but the accumulated phase difference is maintained. Repetition with different
gradient moments of the phase-encoding gradient produces a series of samples
with different phase differences that can be used to calculate the positions.

where t_1 and t_2 are the starting and ending time instants of the phase-encoding gradient, respectively.

This encoding of positions in terms of phase is useful if we collect a series of samples obtained through the repetition of different gradient moments, that is, different gradient values, used to cause different dephasing. The sampling of this phase encoding is performed after the phase gradient has been turned off, during the readout. For this reason, it serves to form the $\mathbf{M}_{xy}$ distribution that is then subjected to frequency encoding.

Provided the gradient directions of phase and frequency encoding are orthogonal, their position encoding effects are also orthogonal functions and so can be used in tandem to provide position encoding in 2 or 3 directions. In fact, using frequency encoding and two phase-encoding directions, it is possible, but time consuming, to generate images of the entire volume within the magnet. A slice selection gradient (discussed in the next subsection) can be useful to reduce acquisition time.

2.2.3   Slice Selection

The volume of the body inside the scanner is often much larger than the region of interest for which images are desired. An RF pulse at the Larmor frequency will affect much of what is in the magnet. Much more efficient is to collect the image data from a selected slab. This can be achieved by slice selection, the simultaneous application of an FR pulse (of proper shape modulation) and a gradient.

Selective excitation allows the excitation of a well-defined slice of the sample. The technique is based on amplitude-modulated RF pulses applied during a gradient is on. With a slice selection gradient applied, the bandwidth of the sample is increased and the RF pulse is able to excite just a slice of the sample. The magnetic field gradient will act as a selector of a part of the sample to be excited: a well-defined region of the sample, the selected slice, will be associated to the RF frequency bandwidth (see Figure 2.20). The thickness of the slice can be controlled as can the position so that a series of slices can be excited and imaged separately (2D imaging) or a thicker slab excited and slices reconstructed based on phase encoding. The magnetic field gradient allows the correspondence between these frequencies and the spatial positions along the gradient direction. The choice of a well-determined gradient value implies that some parts of the sample, having resonance frequencies external to the RF pulse bandwidth, will remain unexcited. As an example, suppose we want to select a slice on the xy plane: in this case, the slice selection gradient has to be directed along z, $\mathbf{G}_z$, perpendicularly to the given plane. The resonance frequency will be function of the position along z. The thickness of the slice, Δz, including the spins to be excited, is determined by the bandwidth $\Delta \omega$ of

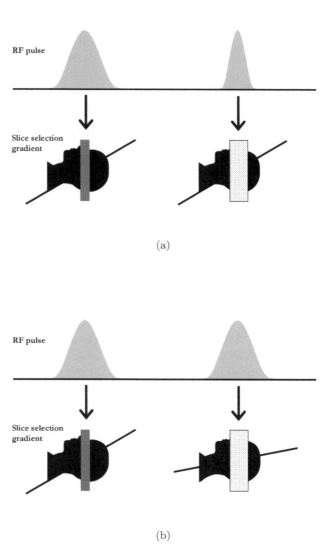

(a)

(b)

Figure 2.20

The presence of a gradient during an RF pulse produces a selection of a slice whose thickness depends on the bandwidth of the RF pulse. (a) A shorter RF pulse produces an enlargement of the slice thickness. (b) The same result can be obtained by reducing the gradient amplitude, while RF duration is maintained unchanged.

the RF excitation pulse and by the $\mathbf{G}_z$ amplitude as follows:

$$\Delta z = \frac{\Delta \omega}{\gamma G_z}.$$ (2.13)

The relationship relating the RF pulse duration and shape to the duration and shape of its frequency spectrum is also a FT relationship as has been discussed in Chapter 1. The aspect of the FR pulse determining the thickness of the selected slice is the RF pulse bandwidth which is determined by the RF pulse duration: having fixed the gradient value, to a shorter RF pulse duration will correspond a wider slice (see Figure 2.20(a)). Equally, having fixed the RF pulse duration, a decrease of the gradient would lead to a wider slice (see Figure 2.20(b)). The position of the selected slice along z can be adjusted by changing the central frequency of the FR pulse (Figure 2.21). The profile of the selected slice corresponds, approximately, to the shape of

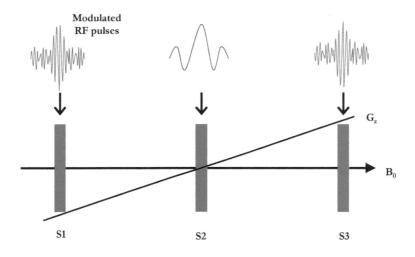

Figure 2.21
RF pulse amplitude modulation allows the position of a selected slice to be controlled. A reduction of the central frequency results in a left shift and an increasing of the central frequency results in a right shift. This is performed in presence of the slice selection gradient, without changing its value.

the RF pulse spectrum. The simple "on-off" RF pulse will have an infinite bandwidth and, for this reason, it does not allow a clear definition of the slice.

To improve slice definition, it is necessary to modulate the amplitude of the FR pulse. The functions used to modulate the RF pulse are often the *sinc* or the *Gaussian* shapes: the *sinc* shape will furnish a better approximation of a rectangular slice shape because the FT of a *sinc* is a square pulse (a *rect*), as needed for a slice profile (see Chapter 1), though truncation of the FR pulse will still lead to an imperfect slice profile.

2.3 Conventional Imaging Techniques

Image reconstruction in MRI is closely related with the data collection method. The most commonly used acquisition techniques are *Spin Warp* (also called Fourier imaging) and *Projection imaging*. Before describing these imaging techniques, it is important to remember that the time-varying signals can be analyzed by following trajectories evolving with time in a 2D or 3D space. This corresponds to a domain, that is, Fourier conjugate of the standard spatial domain containing the object magnetization [71]. For this reason, a FT is used to obtain a depiction of the object's spatial distribution from the collected signals.

The Fourier transform domain is conventionally called k-space, and the conjugate variable of position denoted by the letter k with a subscript (e.g., k_y).

In the following, we describe both spin warp and projection imaging as they relate to k-space for planar reconstruction, that is, two-dimensional (2D) images. The 3D case can be explained as an extension of the 2D case, and it is not discussed here.

2.3.1 Spin Warp Imaging

Most MR images are collected using spin warp imaging, as developed in Aberdeen (Scotland, UK) at the end of 1970s [30]. This method combines the three gradients applied along mutually orthogonal directions for slice selection, phase, and frequency encoding.

The steps in a spin warp, an example of a spin-echo imaging sequence is sketched in Figure 2.22, requires the usage of the following gradients:

1. A slice selection gradient, by convention along the z direction while an RF excitation pulse is applied, to define the image plane

2. A phase-encoding gradient pulse applied along the y direction and positioned between the RF excitation pulse and the acquisition interval

3. A frequency-encoding gradient applied, along the x direction, during the acquisition interval

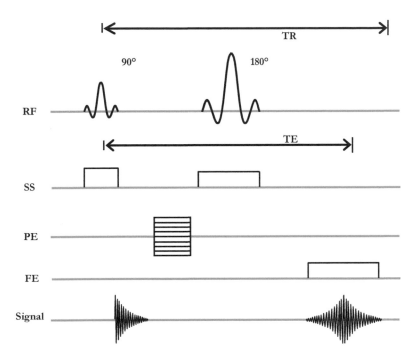

Figure 2.22
Spin echo imaging sequence timing: gradients have been added. For simplicity, negative lobes of slice selection and frequency-encoding gradients, used to maintain phase coherence, have been omitted.

This method for image acquisition/reconstruction allows to the class of the so-called 2D-FT, or two-dimensional imaging techniques that use FT for image reconstruction.

The central idea for the 2D-FT techniques is that a one-dimensional (1D) projection of the sample is obtained by FT of the signal evolving with time in presence of a frequency-encoding magnetic field gradient. To extend this technique to the 2D image reconstruction, it is necessary to codify the signal evolution as a function of a second "dimension," indicated as "pseudotime," through the application of a phase-encoding gradient orthogonal to the slice and frequency-encoding directions.

The signal temporal evolution in presence of the frequency-encoding gradient, also called *readout gradient* when the signal is collected during its application, is repeated many times (instances or readouts): at each iteration, the phase-encoding gradient amplitude is incremented by a fixed amount to produce signal evolution in pseudotime that relates to position in the phase-encoded direction. The effect of the spin-warp imaging sequence is sketched in Figure 2.23. In particular, we consider a spin packet (a voxel) in the point whose coordinates are (x, y) allowing to the selected slice. The readout gradient, $\mathbf{G}_x$, introduces an accrued phase, which depends on time t, position x and that, using Equation 2.12, is given by:

$$\phi_1(t, x) = \gamma x(t - t_0)G_x \tag{2.14}$$

where t_0 is the instant at which the gradient is turned on. On the contrary, the phase-encoding gradient $\mathbf{G}_y$ produces a loss of phase that depends on the position y and on the pseudotime n as:

$$\phi_2(n, y) = \gamma y(t_2 - t_1)G_y(n) \tag{2.15}$$

t_1 and t_2 being the times where $\mathbf{G}_y$ has been turned on and off, respectively. The gradient intensity, $G_y(n)$, depends on the readout number n of the sequence:

$$G_y(n) = G_0\left(\frac{n - N}{N}\right) \tag{2.16}$$

where G_0 is the maximum phase encoding gradient amplitude and the integer n varies between 0 and $2N$, given $2N+1$ the total phase-encoding gradient values, having considered its symmetry around 0.

The signal generated by the voxel (x, y) will depend on the real time t and on the pseudotime n. The final image is obtained as a 2D-FT of the collected signals ordered into a table whose axes are time and pseudotime (for example, columns and rows, respectively).

A variation of spin-echo imaging sequence, the gradient-echo sequence, can be implemented by replacing the second RF pulse (180°) in Figure 2.22 with a negative gradient pulse applied along the readout direction.

3D imaging can be obtained either by successively imaging adjacent slices (through RF amplitude pulse modulation changes and a slice selection gradient) or by using a further phase-encoding gradient perpendicularly to the other phase-encoding and frequency-encoding gradients.

2.3.2 Imaging from Projections

This method was the first used by Lauterbur when invented MRI [66] and derives directly from Computerized Tomography.

To collect data for the reconstruction of a 2D image, in the xy plane, a slice selection gradient is first applied along the z direction during which an RF excitation pulse is played out. After slice selection, spatial encoding is obtained by applying a second gradient G along a given direction ϕ in the xy plane.

The FT of the collected signal represents the projection, $p_\phi(r)$, of the slice

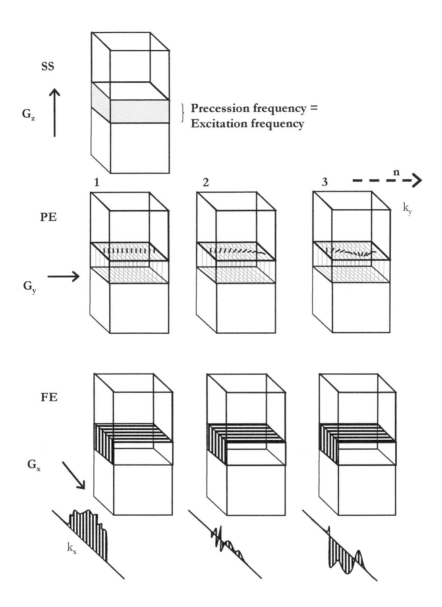

Figure 2.23
The role of the gradients in a spin echo imaging sequence.

$f(x, y)$ at the angle ϕ, and is related to the desired image by the following relationship:

$$p_\phi(r) = \int_{-\infty}^{\infty} f(r, s) ds \tag{2.17}$$

where (r, s) represents an orthogonal cartesian system rotated to the angle ϕ with respect to the original (x, y) system, as shown in Figure 2.24, such that:

$$
\begin{aligned}
r &= x \cos \phi + y \sin \phi \\
s &= x \sin \phi - y \cos \phi.
\end{aligned}
\tag{2.18}
$$

Complete determination of the 2D image $f(x, y)$ can be achieved by mea-

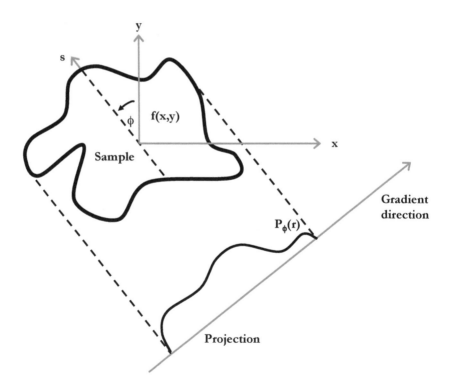

Figure 2.24
A projection at an angle ϕ of a 2D sample is reported. Mathematically, it represents the integral of the sample along the direction perpendicular to that of the projection.

surement of a set of projections on the xy plane, by changing ϕ. In order to do this, the intensities of the gradients are varied as follows:

$$
\begin{aligned}
G_x &= G\cos\phi \\
G_y &= G\sin\phi.
\end{aligned}
$$

(2.19)

with angle ϕ, indicating the direction of the resulting gradient, being varied between $0°$ and $180°$

Reconstruction of the image, from the measured projections is then achieved by mathematical algorithms such as Fourier reconstruction [83] or Filtered Back Projection [28].

Although it has been demonstrated that Fourier reconstruction and Filtered Back Projection give equivalent results, the first has been demonstrated to be computationally more efficient [102]. For this reason, we briefly introduce Fourier reconstruction (for a deeper insight into these reconstruction methods, please refer to [83, 28]).

The Fourier reconstruction method depends, fundamentally, on the *projection slice theorem* [83], stating that each Fourier coefficient of the spin density function (the desired image) is equal to the corresponding Fourier coefficient of the projection passing through (crossing) that point. Following from this theorem, each collected projection represents a *line* through the 2D FT of the sampled image (in 3D MRI, the theorem also holds and the collected projection represents a *slice* through the 3D FT of the sampled volume).

Thus, starting from a set of projections, the Fourier reconstruction method can be described as follows:

1. Place each 1D projection on the 2D k-space plane at the proper angle and position

2. Interpolate the collected data (distributed on concentric circles) to obtain a Cartesian matrix (this operation is often referred as *regridding*)

3. Perform the 2D FT to obtain the sample image

Step 2 is very important for Fourier reconstruction and different interpolation methods can be used. A simple linear interpolation method yields good results in most MRI applications, when a large number of projections is used, though other interpolation methods can be used [83, 104].

3D reconstruction can be obtained either by repeating imaging reconstruction from adjacent slices (through RF amplitude pulse modulation changes), by using a phase-encoding gradient along the third direction, or by generalizing the projection to include a gradient component in the 3D spatial direction.

If this last option is chosen, a second angle θ must be defined to take into account of the tilting with respect to the z axis (see Figure 2.25).

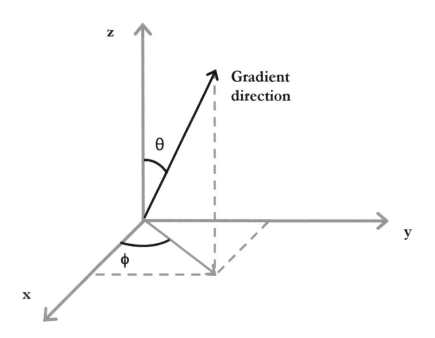

Figure 2.25
The projection direction is chosen by using a combination of gradients, as reported in Equation 2.20, following two angles: the first, ϕ, indicates the direction on the xy plane; the second, θ, is defined to take into account the tilt angle with respect to the z axis.

In this case, Equations. 2.19 are modified as follows:

$$G_x = G\cos\phi\cos\theta$$
$$G_y = G\sin\phi, \cos\theta$$
$$G_z = G\sin\theta.$$

$$(2.20)$$

A feature of imaging from projections is that it can be performed either by using a direct combination of readout gradients, by collecting the FIDs, or by using refocusing pulses (RF or gradient pulses) to collect echoes. Difference between them is that the former allows a direct signal measurement, immediately following the 90° RF pulse (in practice, a brief gap is required to allow switching from transmission and reception and correct for dephasing due to slice selection gradient); the latter requires an additional pulse before the echo is collected.

The former has the following advantages:

1. Sampling the FID in this way avoids the effects of T_2 and T_2^* decay, thus providing a maximal signal amplitude.

2. The sampling starts at the centre of the Fourier image domain (i.e. where the FID is maximum) and proceeds outward as a ray: motion occurring during signal detection more probably will affect the outer part (lower amplitude part) of the Fourier domain.

The latter has the advantage that the resulting SNR is better ($\sqrt{2}$ times greater) for echo reconstruction than for FID reconstruction, due to the fact that with echo we collect twice the information required to obtain a projection than with a FID.

On balance the former is preferred, though two problems have to be solved.

The first is that a method to recombine half Fourier data [34, 75, 132] has to be used to reproduce, from a FID (see Figure 2.11), the classical "bell" shape of an echo. Alternately, a smoothing filter has to be multiplied to the FID to attenuate its abrupt starting jump. These operations are necessary to avoid the modulation effects induced by this jump, not for imaging purposes.[2]

The second is that it could be necessary to discard some initial points of the FID to avoid interference with a residual signal coming from the transmitter because its turning off is not instantaneous.

From now on, when discussing acquisition from projections, whether FIDs or echoes were collected, we assume both negative and positive frequencies are collected because, in principle, a FID is sufficient to give information about the whole projection.

---

[2]The FID shape resembles the multiplication between a nonperiodic square pulse and the oscillating signal. For this reason, the FT of the FID corresponds to the convolution between the *sinc* function (see Figure 1.2) with the FT of the oscillating signal. This convolution strongly distorts the signal if an half Fourier recombination or a smoothing filter are not used.

After its first application in MRI, imaging from projections was largely abandoned in favor of Fourier imaging because the latter allowed faster reconstruction.

In the last years, imaging from projections has been reemerging because of its low sensitivity to motion artifacts and great versatility [105].

2.4   Bandwidth, Sampling, Resolution, and Sensitivity

In MRI, some parameters are fixed, while others are under operator control.

The field strength, maximum gradient strength, and gradient slew rate for example are fixed by manufacturer. The choice between sequences, sequence timing, matrix size, slice thickness, gap between slices, FOV, number of excitations, orientation of imaging planes, type of receiver coil, use of gating, use of contrast, etc., are under operator control.

An important imaging parameter is represented by the FOV, the minimal spatial region (usually a square in 2D, or a parallelepiped in 3D) occupied by the image. In the frequency-encoding direction, x, the FOV extent L_x (Figure 2.26) is determined by the sampling rate, that is the signal sampling frequency, and by the strength of the gradients, that is by the range of precession frequencies (the "readout bandwidth") in the frequency encoding direction. The sampling rate, indicated as $2B_\nu$ ($\pm B_\nu$), is related to the distance in time Δt between consecutive samples (known as *dwell time*) as follows:

$$2B_\nu = \frac{1}{\Delta t}. \qquad (2.21)$$

Along x, the full bandwidth across the FOV is $\gamma G_x L_x$ and the FOV extent is

$$L_x = \frac{2B_\nu}{\gamma G_x}. \qquad (2.22)$$

Another important factor is D the object extent along x (Figure 2.26) necessary to define the bandwidth of the object (the "signal bandwidth") defined as $\gamma G_x D$ (according to the readout bandwidth, the signal bandwidth can be indicated as $2S_{B_\nu}$, $\pm S_{B_\nu}$).

Though the object extent is not an imaging parameter, imaging result can be influenced by the relationship between the object extent and the FOV extent. In fact, in keeping with the discussion of the Nyquist theorem and aliasing in Chapter 1, if the signal is sampled at Δt, the FT of the sampled signal is replicated at $1/\Delta t$ (i.e., the sampling rate and so the FOV). It is important that the bandwidth contained in the signal remains lower than, or equal to, the sampling rate (or equivalently, that the object does not extent beyond the FOV) to avoid aliasing.

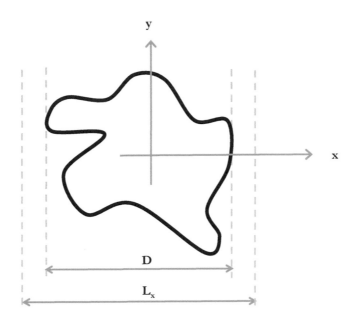

Figure 2.26
The extent of the FOV along x, L_x, is necessary to calculate the acquisition parameters. The extent of the object along x, D, can influence imaging results.

This result can be directly applied to the readout process in MRI by calculating the bandwidth of the signal resulting from spin precession in the presence of the readout gradient. In the frequency-encoding process, precession frequency is used to recover the spin location.

If the desired FOV L_x is smaller than D (i.e., the $B_\nu \leq S_{B\nu}$), the signal bandwidth must be reduced. This can be done by applying a band limiting filter to the receiver (also called an analog *antialiasing* filter or hardware filter) prior to the sampling step.[3]

This operation can also be also be performed by increasing the amplitude of the readout gradient value and, proportionally, the FOV bandwidth to make the readout bandwidth equal to the whole receiver bandwidth (in this case, the receiver acts as an hardware antialiasing filter). This is possible by

---

[3]It is worth noting that with phase encoding, the signal coming from outside the FOV cannot be eliminated by an hardware filter and some aliasing has to be tolerated.

supposing to increase the gradient of what we need. As will be clarified at the end of this section, this could be not always possible.

A third option is to reduce the signal bandwidth, by reducing the readout gradient, while maintaining the same dwell time (the readout bandwidth is leaved unchanged). In this case, the object is "restricted" to fit a lower region: some small details can be lost (i.e., the spatial resolution[4] is lowered).

Nyquist sampling requirements apply both to the k-space and spatial domains, and the temporal sampling requirement is related to the k-space sampling requirement. Assuming a constant readout gradient, the interval between readout points in k-space Δk_x is:

$$\Delta k_x = \gamma G_x \Delta t. \tag{2.23}$$

Combining Equation 2.23 with Equations 2.21 and 2.22, we obtain the k-space Nyquist requirement:

$$\Delta k_x = \frac{1}{L_x}. \tag{2.24}$$

If we collect N_x readout points, the maximum k-space extension is

$$N_x \Delta k_x = \frac{N_x}{L_x} = \frac{1}{\Delta x} \tag{2.25}$$

that is,

$$\Delta x = \frac{1}{N_x \Delta k_x} = \frac{1}{N_x \Delta t \gamma G_x} \tag{2.26}$$

where Δx is the image spatial resolution in the readout direction, that is, the pixel size after the FT. Equations 2.24 and 2.26 state that the k-space resolution, Δk_x, is the inverse of the length of FOV and that the spatial resolution, Δx, is the inverse of the extent of k-space, respectively.

The last part of the Equation 2.26 summarises the relationship between spatial resolution, gradient intensity, and readout interval. Some attention is warranted when applying Equation 2.26, as a literal application of the equation seems to allow us to indefinitely increase spatial resolution (reducing Δx) by increasing indefinitely the readout interval or gradient amplitude. Theoretically, this is true, but in practice false. In fact, Δx can be nominally reduced as we want, but there is a minimum value that cannot be exceeded (it is impossible to represent different structures whose distance is lower than this threshold) due to the decaying nature of the MRI signal and the limited sensitivity of the MRI technique/equipment. The instrument is unable to register signal variations below some sensitivity threshold. In other words, above a certain value, the increment of the readout interval has the same effect as we pad (numerically) the same interval with zeros.

The sensitivity depends on SNR considerations; the voxel must contain

---

[4]Spatial resolution refers to the smallest resolvable distance between two different objects, or two different features of the same object [40].

enough spins to be detectable by the MRI receiver system. For a given image with three isotropic spatial resolution Δx, the voxel is a cube with volume Δx^3. If s is the minimum number of spins that can be detected by the spectrometer and ρ is the mean spin density in the sample, then $\Delta x^3 \rho \geq s$ must hold. This gives

$$\Delta x \geq \left(\frac{s}{\rho}\right)^{1/3} \tag{2.27}$$

as the minimum value for Δx and represents a threshold for spatial resolution. It is impossible to go below that value and, if we use too-high imaging parameters values, such as the gradient value or the readout interval, the resulting image will have reduced SNR without any increasing of useful information and so image quality will deteriorate.

Practical aspects such as the maximum usable gradient strength of the system and the system memory for the readout and acquisition time may further modify the resolution that can be achieved.

2.5 An MRI Scanner

MRI requires the generation of highly uniform and stable magnetic fields in the imaging region to produce the magnetization vector. The standard figure of merit for expressing field uniformity is the maximum field variation referred to the field value in the center of the imaging region, calculated over the imaging region itself, the volume of interest (VOI); this value is usually expressed in ppm (parts per million). Typical field uniformity for whole body MRI systems varies from 10 - 20 ppm (relative homogeneity) for low-field scanners (about 0.2T), based on permanent magnets, to 1 - 2 ppm for higher-field scanners (about 2T), based on superconducting magnets. The size and design of the magnet will determine the size, shape, and location of the homogeneous volume usable for imaging. Inhomogeneity in the main magnetic field tends to result in image distortions.

In addition, an imaging system requires magnetic field gradients that can be switched on and off rapidly to produce the intentional magnetic field variations across the sample necessary for spatial localization, as discussed above.

Moreover, radio transmitters/receivers are necessary to transmit/receive RF pulse to/from the sample. Summarizing, the essential components of an MR imaging system (see Figure 2.27) include

1. A large magnet which generates a uniform magnetic field.

2. Smaller electromagnetic coils to generate magnetic field gradients for imaging.

3. A radio transmitter and receiver and its associated transmitting and receiving coils.

4. A computer to coordinate data acquisition, image reconstruction, and display. Imaging sequences are implemented in software.

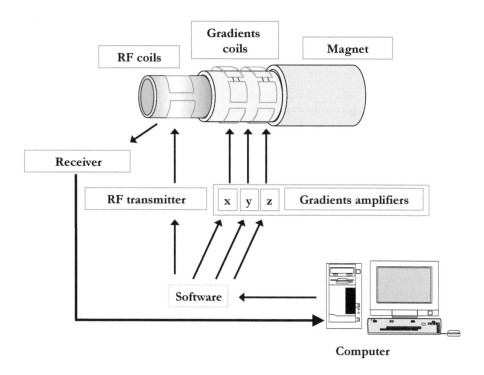

Figure 2.27
An MRI apparatus is composed by the following components: a magnet to generate the main magnetic field, a set of gradient coils, RF transmitter receiver coils, a personal computer to drive the acquisition process and to perform image reconstruction and presentation.

2.6   Bloch Equations and Numerical MRI Simulators

The physical MRI system is fundamental to collecting real data from a sample but, when the scope is to study and test innovative acquisition sequences or to simulate the effects of modification of some parameters or of some external disturbances on the produced images, it can be useful to study them on a numerical MRI simulator before their implementation on a real experimental apparatus. The main reasons are easiness and rapidness of prototyping, implementation, rapid parameters definition, and optimization; rapid modification; hardware preservation and time-machine saving.

A numerical MRI simulator consists of a computer program implementing the numerical solution of Equations 2.28, which describes the nuclear magnetization $\mathbf{M} = (\mathbf{M}_x, \mathbf{M}_y, \mathbf{M}_z)$ as a function of time in presence of relaxation, according to the Bloch equations [11]:

$$\frac{\partial \mathbf{M}_x(t)}{\partial t} = \gamma (\mathbf{M}(t) \times \mathbf{B}(t))_x - \frac{\mathbf{M}_x(t)}{T_2}$$

$$\frac{\partial \mathbf{M}_y(t)}{\partial t} = \gamma (\mathbf{M}(t) \times \mathbf{B}(t))_y - \frac{\mathbf{M}_y(t)}{T_2}$$

$$\frac{\partial \mathbf{M}_z(t)}{\partial t} = \gamma (\mathbf{M}(t) \times \mathbf{B}(t))_z - \frac{\mathbf{M}_z(t) - \mathbf{M}_{z,eq}}{T_1}$$

$$(2.28)$$

where γ is the gyromagnetic ratio and $\mathbf{B}(t) = (\mathbf{B}_x(t), \mathbf{B}_y(t), \mathbf{B}_0 + \mathbf{B}_z(t))$ is the magnetic flux density experienced by the nuclei.

The z component of the magnetic flux density $\mathbf{B}$ is typically composed of two terms: one, $\mathbf{B}_0$, is constant in time (the main magnetic field); the other one, $\mathbf{B}_z(t)$, is time dependent (determined by the gradients applied). $\mathbf{M}(t) \times \mathbf{B}(t)$ is the cross product (vectorial product) of these two vectors.

Most of the results reported in the following chapters of this book were obtained using an MRI simulator because in many cases the corresponding pulse sequences have not been implemented. The flowchart of the simulator used [106], implemented in Matlab [133], is shown in Figure 2.28.

The numerical sample, that is passed to the algorithm, consists of a 4D representation of a sample 3D image in which each voxel has associated a list containing values of the spatial position of its center (x, y, and z values), its proton density, relaxation times (T_1 and T_2 values), and local stationary magnetic field value (to account for static magnetic field inhomogeneities, thus considering T_2^*) as well as the chemical shift (in ppm).

The pulse (or imaging) sequence is defined as a table of temporal state changes (reported on the rows) and six columns: one indicates elapsing time, three reporting the gradients amplitudes (G_x, G_y, and G_z), one the RF pulse, and one the number of the acquired sample. These values are determined on

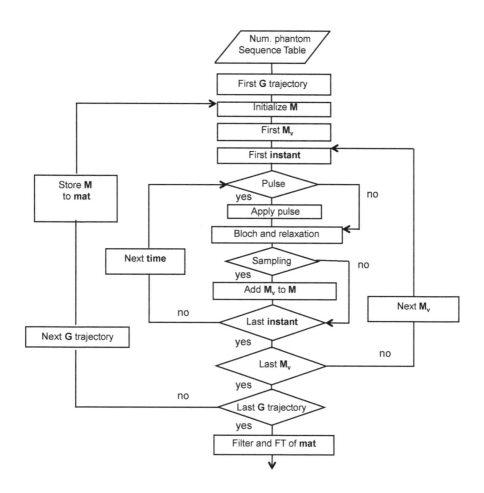

Figure 2.28
Flowchart of an MRI numerical simulator.

the basis of the RF pulses (number of pulses, application time, duration, and shape), sampling time, sampling frequency, repetition time, gradient amplitude, and gradient orientation ($\mathbf{G}$) specified by the user.

The signal collected by an MRI scanner derives from the global magnetization of an excited slice of the sample. The algorithm cannot handle the complex nature of the global magnetization directly, but rather operates on the net magnetization vector of one voxel at a time. The algorithm describes the temporal evolution (**time**) of each $\mathbf{M}_v$ subject to the events of the sequence table, modified according to Equations 2.28, following the application of RF or gradient pulses (Pulse) and evolves in the course of a given pulse sequence (a series of events). The magnetization vector from one voxel of the simulated object is denoted $\mathbf{M}_v$. The sum of all $\mathbf{M}_v$ gives the global magnetization vector ($\mathbf{M}$) for the current gradient setting $\mathbf{G}$.

The application of a Pulse serves to modify the $\mathbf{M}_v$ direction. The Bloch equations serve to describe the rotational phenomena occurring in the $\mathbf{M}_v$, following the application of perturbing pulses. They describe the components of $\mathbf{M}_v$ as a function of time. The relaxation processes affecting $\mathbf{M}_v$ are calculated according to the same equations, with the inclusion of the relaxation functions (Bloch and relaxation).

For each instant of the acquisition process, the x and y components of $\mathbf{M}_v$ (those perpendicular to the main magnetic field direction) are added to the current contents of $\mathbf{M}$ in that instant (see the instruction "Add $\mathbf{M}_v$ to $\mathbf{M}$" of the flowchart).

A linked list of $\mathbf{M}$ values stores all the sampling instants. When this operation has been repeated for each $\mathbf{M}_v$ at all instants, the list contains the temporal evolution of the global magnetization vector components providing simulated dataset for the signal collected from the target object using the given acquisition sequence. The control of gradient values describes a specific well-defined trajectory in the k-space.

When the whole list is complete, the content of $\mathbf{M}$ is stored into a matrix (**mat**), $\mathbf{M}$ is reinitialized and the gradient setting is updated. The process is then repeated to collect another k-space trajectory. The acquisition ends when the target k-space coverage has been obtained.

The resulting image is reconstructed by 2D FFT of the **mat** coefficients (2D FFT of **mat**) after they have been low-pass filtered. The 3D reconstruction of the sample can be obtained by repeating the application of the algorithm to consecutive planes of voxels.

Part II

Limitations of
Conventional MRI

3

Limiting Artifacts for Advanced Applications

CONTENTS

Artifacts appear in MRI for a variety of reasons. They appear as signals or voids in the images that do not have an anatomic basis, or are the result of distortion, addition, or deletion of information. They can degrade images sufficiently to cause inaccurate diagnosis. Many MR artifacts are neither obvious nor understandable from previous experience with conventional imaging MRI. While some artifacts are scanner specific, the majority are inherent in the physics or method of MRI imaging. For this reason, books dealing with MRI often contain chapters devoted to most common artifacts in MRI and to their sources. With a deep understanding of their cause, most MRI artifacts can be corrected, minimized, or avoided, through appropriate action during data acquisition or the reconstruction stage. This requires familiarity with scanner design, physical processes, and image acquisition.

Most common sources of artifacts are magnetic field inhomogeneity, motion, and undersampling. The first occurs because, although magnet construction and shimming are very accurate, residual inhomogeneity remains (or is introduced by the object being studied) leading to geometrical distortions in the images. The last two are strongly connected: a long acquisition time ensures complete sampling and, in principle, high image quality but frequently it produces artifacts due to sample motion during the examination (voluntary or involuntary movements may occur). Conversely, reducing the acquisition time may diminish the amount of motion that occurs but is often obtained by incomplete sampling (some k-space trajectories are skipped), thus producing undersampling with its associated artifacts.

In particular, ultra-fast applications, such as functional or real-time imaging, require us to deal with undersampling. This emerges because, although the effects of some motions (e.g., respiration and cardiac pulsation) can be greatly reduced through triggering or gating, there remains some motion over the course of scan acquisition, while the effects of irregular motion such as peristalsis persists. These latter can be diminished by speeding up the acquisition in the hope of reducing the probability that the motion occurs during scanning.

In what follows we deal with these specific artifacts, providing detailed description of their physical background and manifestations. The arguments treated serves as motivation for the development of unconventional acquisition/reconstruction methods, introduced in the following chapters, that can be used to overcome these problems.

3.1   Magnetic Field Inhomogeneity

Any phenomenon disturbing the linear relationship between frequencies and spatial positions violates the assumptions of MRI and induces errors in the image.

Inhomogeneity of the main B_0 field [58], magnetic susceptibility [120], and chemical shift [5, 89] are typical sources of such inconsistencies in the resonant frequency–position relationship. Due to the fact that the negative effects are very similar, we treat just B_0 field residual inhomogeneity.

It should be noted, however, that magnetic susceptibility or chemical shift can be considered a nuisance to one application and a boon to another. For instance, the differences in the magnetizability, or susceptibility, of close materials may lead to displacement, blurring, or signal drop-out artifacts; yet, the susceptibility difference induced by the oxygenation of hemoglobin is the foundation of the blood oxygen level dependent signal in functional MRI studies [15].

Analogously, chemical shift can produce displacement artifacts in morphological imaging but also provide valuable spectral information about metabolite content in an extended region and so adds chemical analysis of body tissues to the clinical utility of magnetic resonance [55].

The B_0 field is never perfectly uniform throughout the magnet. Residual inhomogeneity causes signal amplitude modulation that the reconstruction process assumes are due to the prescribed spatial encoding process. The spatial localization/encoding in zones affected by inhomogeneity is inconsistent with the actual spatial position: the signal coming from one district appears to be coming from another, thus distorting the reconstructed image. If, for example, the inhomogeneity causes three points to experience the same frequency during readout, their signal will be superposed on one point on reconstruction, as shown in Figure 3.1. Even knowing which three points were encoded in this way is not sufficient to restore the signal to its proper distribution because one does not know the fractions of the superposed signal to allocate to each site.

This information can be obtained by repeating the acquisition process with

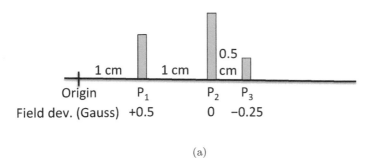

(a)

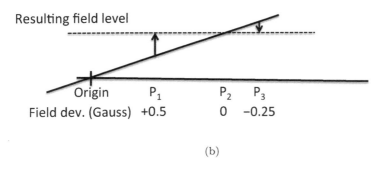

(b)

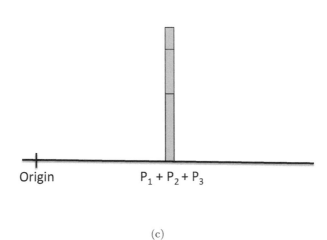

(c)

Figure 3.1
Different positions (P_1, P_2, and P_3) have local residual inhomogeneity (a). When the gradient is on, this yields equivalent net magnetic fields (b), corresponding to signals with the same precessional frequencies during sampling: the three positions appear in the image at the same point (c).

different gradient values resulting in a long process that can also lead to an ill-conditioned problem if the experimental noise is high (i.e., data have low SNR) [60].

The effect of the image distortion is reported in Figure 3.2. In particular, Figure 3.2(a) illustrates an image unaffected by magnetic inhomogeneity, while Figure 3.2(b) shows the same object imaged in presence of a residual $\mathbf{B}_0$ inhomogeneity, where alterations in the image geometry, shape, and intensity are evident (some differences in image intensities can also be due to some change in acquisition parameters). The severity of deformation is proportional to the residual inhomogeneity. These artifacts can then be eliminated by nullifying the residual inhomogeneity; a first step in this direction is to use high-quality magnets. In addition, a long shimming process is necessary to obtain optimized magnetic field homogeneity inside the ROI. Figure 3.3 highlights the difference in magnetic field precision before, Figure 3.3(a), and after, Figure 3.3(b), the shimming procedure. Initially, the magnetic field inhomogeneity is so severe that an accurate imaging would be impossible with conventional techniques. After shimming, only the extreme periphery shows notable inhomogeneity.

It is impossible to eliminate completely residual magnetic field inhomogeneity through shimming. Magnetic field uniformity is often indicated in parts per million (ppm, expressed as $10^6 * \Delta\mathbf{B}_0/\mathbf{B}_0$), though the effects it produces are independent of the main magnetic field value. In most commercial whole body MRI scanners, an acceptable level of inhomogeneity is of the order of 30 ppm. This level of uniformity can be tolerated in standard morphological images. For a magnetic field of 1.5 T (1 Tesla corresponds to 10000 Gauss), a precision of 30 ppm would correspond to 450 mGauss. Obviously, the effects of a residual inhomogeneity are strongly dependent on its distribution. In fact, if an error of 450 mGauss is distributed over a ROI of 20 cm, corresponding to a dedicated magnet, the distortion in the resulting image would be more evident than the same error distributed over a ROI of 45 cm, which is often the case of a whole-body scanner.

The requirement for traditional MRI that the static magnetic field, $\mathbf{B}_0$, must be as homogeneous as possible has significant effects on the design, the shape complexity, and the usability of the scanners. Magnet cost grows exponentially with the required magnetic field uniformity, thus affecting the price of systems. Moreover, the resulting magnets are often narrow and uncomfortable tubes.

One can imagine a variety of situations where it might be useful to perform imaging with the sample placed entirely outside the magnet's bore, that is, to perform imaging from "open" magnets (see Figure 3.4).

Some of the applications that would benefit include imaging during surgical interventions (also called "interventional" imaging), surface imaging (for

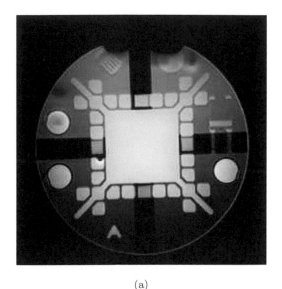

(a)

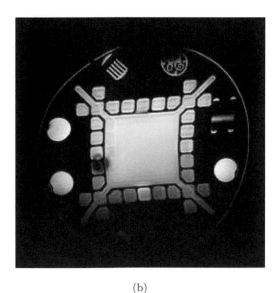

(b)

Figure 3.2
Images illustrating relationship of distortion due to inhomogeneity in the magnetic field with the echo time TE: (a) an undistorted image acquired at short TE; (b) at long TE, the global
$\mathbf{B}_0$ inhomogeneity becomes evident. Both images show the effects
of an intrinsic inhomogeneity due to a bubble. Reproduced from
http://www.ccn.ucla.edu/bmcweb/sharedcode/MRArtifacts/MRArtifacts.html
(authorship and copyrights: UCLA Brain Mapping Center).

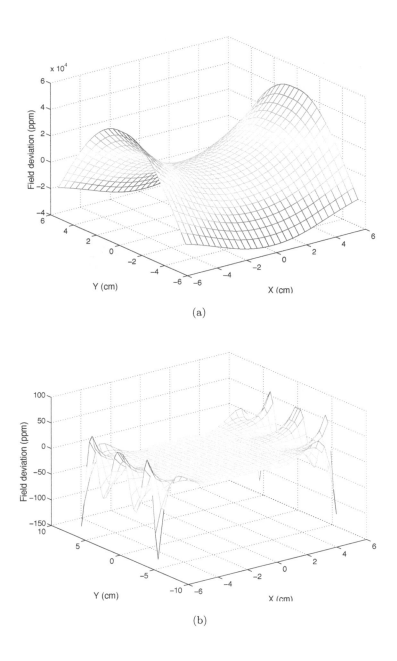

(a)

(b)

Figure 3.3
Residual magnetic field inhomogeneity for a dedicated permanent magnet: (a) before shimming, (b) after shimming.

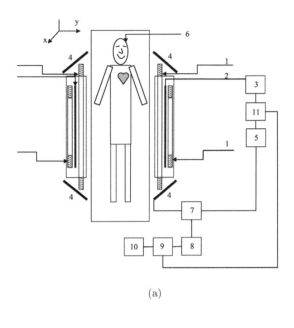

(a)

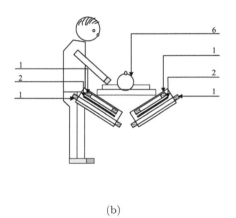

(b)

Figure 3.4
Different views of a theoretical, completely open magnet: (a) bird's-eye view;
(b) head-feet view where an operator is represented to illustrate the ease of
intervention on the patient. Numbers refer to single parts of the system: (1) the
magnet; (2) the gradients; (3) gradients power supplies; (4) RF transmitter; (5)
mixer; (6) the patient; (7) RF switch; (8) demodulation unit; (9) signal/image
processing unit; (10) display; (11) sequence control unit.

example, of skin lesions), imaging during movement (of arms, legs, neck, or back), imaging of obese or claustrophobic people, as well as children or those suffering from neurological or psychiatric diseases. In open magnets, meeting the ppm magnetic field precision required for traditional MRI methods is very difficult.

It is worth noting that geometrical distortions cannot be eliminated with a posteriori numerical processing: it is possible to measure the magnetic field residual inhomogeneity distribution and so estimate which positions are degenerate to each other, but to reconstruct the intensities at the distinct points would require knowing also their relative contributions to the superposed signal in the image (see Figure 3.1). Some methods that can avoid the superposition during acquisition will be considered in Chapter 4.

Another important point is that in the above we considered only residual main magnetic field inhomogeneity (i.e., related to $\mathbf{B}_0$) and not that due to the magnetic field gradients. In fact, while a main magnetic field of about 30 ppm at 1.5 T is required, a residual error in the magnetic field generated by gradient coils to be up to 3%–4% is generally accepted. This has a series of explanations.

First, as the gradients are much lower than $\mathbf{B}_0$, a maximum error of 4% on the gradients corresponds to some ppm on the main magnetic field. As an example, a 4% of maximum error in a gradient of 0.1 Gauss/cm would result in an absolute error of 0.004 Gauss per centimetre, that is to 0.18 Gauss within a ROI of 45 cm (in the worst case), as opposed to the absolute error of 0.45 Gauss that would result from an inhomogeneity of 30 ppm in a magnetic field of 1.5 T. To obtain the same absolute error, the main magnetic field would need to have a precision of 12 ppm!

Second, magnetic field inhomogeneity produces a T_2^* reduction: this is true both for the main magnetic field and for the magnetic field gradients (having the role of changing the spin precession frequencies in different spatial districts; the magnetic field gradients heavily influence T_2^*). Nevertheless, the effect due to gradients, with respect to T_2^*, can be eliminated by turning off the gradients. This effect can be reversed by changing sign to the gradients, as it happens during a gradient echo. On the contrary, the T_2^* reduction due to residual main magnetic field inhomogeneity is irreversible.

Third, the magnetic field profile produced by the magnet represents an even function (odd terms due to construction defects are usually much lower than the even terms); see for example Figure 3.3. This fact can lead to image folding and must be corrected with accurate shimming. The gradients, on the other hand, produce odd functions (even terms due to construction defects are usually much lower than the odd terms); see, for example, Figure 3.5. This leads to a function that is monotonic, and the resulting distortion can be corrected numerically. For example, if a gradient having the shape reported in Figure 3.5 is used to encode the image of Figure 3.6(a), it results in the geometrically distorted image of Figure 3.6(b), with the effect of stretching

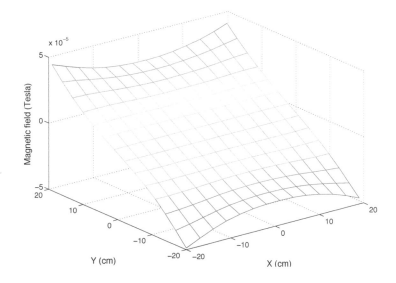

Figure 3.5
An experimental field gradient shape whose maximum inhomogeneity, 4%, is localized to the corners.

the objects situated on the image periphery. An off-line numerical method can be applied in this case to realign the separated zones because overlapping is avoided except for some limited regions where the image was compressed. In the distorted image, bright bands were due to original intensities summed on the collision points, while dark bands (empty zones) were created by signal being separated by their original positions.

The details of some correction algorithms or unconventional acquisition schemes that can contribute to reducing artifacts due to residual inhomogeneity of static magnetic field are described in Chapter 4.

3.2 Motion

Motion-derived artifacts are very common especially in body imaging and can be due to heart or arterial pulsations, breathing, swallowing, peristalsis, tremor, and gross movement of a patient. The artifacts appearance depends on motion typology: a first distinction can be made between random and periodic motion. Random motion during imaging generally results in blurring, whereas periodic motion produces ghost images. As an example, consider the MRI image of Figure 3.7.

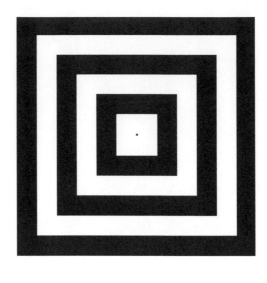

(a)

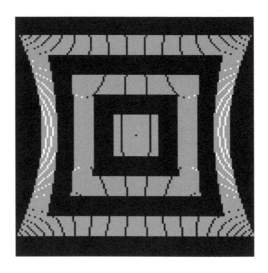

(b)

Figure 3.6
A numerical image composed by alternate black and white squares (a), has been used to verify spatial deformation due to a gradient along the y direction whose shape is that of Figure 3.5. The resulting image is reported in (b).

A second distinction can be made based on when and where the motion occurs during the scan. If motion takes place during data acquisition, the motion's path contributes as a term in the data encoding. However, artifacts are different if motion occur in the middle of the acquisition interval or at the extremities of the acquisition interval. In the first case, low spatial frequency artifacts are generated (Figure 3.7(a)); in the second case, high spatial frequency artifacts are present (Figure 3.7(b)).

On the other hand, motion between acquisition intervals causes inconsistencies: data collected before and after the movement do not correspond to the same anatomical arrangement or placement of the tissues. The frequency content from each arrangement is therefore incomplete, and the image becomes a superposition of different undersampled arrangements with artifacts in the form of aliasing and displacement. The aliased shapes are different for different imaging sequences [128], and can vary from "ghosting" in spin warp imaging to "streaking" or "swirling" in imaging from projections or spiral trajectories. In any case, the resulting images are blurred and show loss of resolution and SNR.

Data inconsistency artifacts, due to periodic motions, can largely be mitigated by using a combination of gating, navigator echoes, motion correction sequences (such as those used for blood saturation), and post-processing techniques [139, 2, 4, 82, 3].

Random motions, on the contrary, can be avoided only by reducing acquisition time.

One way to speed up imaging, when the repetition time for a sequence has been minimized, is to relax the constrain of "completeness" and to accept "undersampling," that is, to avoid the measurement of some of the k-space trajectories. Imaging time reduction can be extremely useful for diagnostic and clinical purposes.

3.3 Undersampling

Recently, MRI has developed considerably in the direction of dynamic imaging opening up several new applications such as monitoring of contrast agent dynamics [117, 147, 118, 148], mapping of human brain neural activity based on blood oxygenation level-dependent (BOLD) imaging contrast [65], MR-guidance of biopsies [72, 69, 17], monitoring of ablations [81, 22], guidance procedures [64, 36], and real-time visualization of cardiac motion [63, 115]. Although these developments are generally promising, their application is limited by the compromise between temporal and spatial resolution. In conventional MRI, the number of collected data points in a fully sampled k-space set is

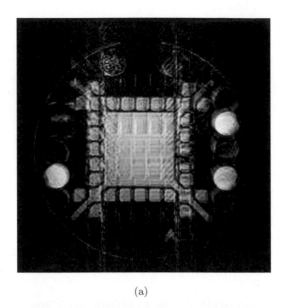

(a)

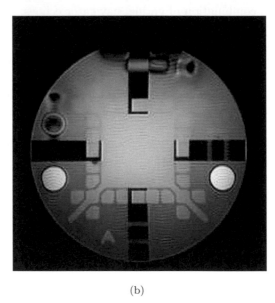

(b)

Figure 3.7
Image artifacts due to motion: (a) motion occurring in the middle of the
acquisition interval; (b) motion occurring at the extremities of the acquisition interval. The undistorted image is that of Figure 3.2(a). Reproduced from
http://www.ccn.ucla.edu/bmcweb/sharedcode/MRArtifacts/MRArtifacts.html
(authorship and copyrights: UCLA Brain Mapping Center).

determined by spatial resolution requirements and the Nyquist criterion for the alias-free field of view.

To improve temporal resolution many approaches use "undersampling" [115, 80, 131, 119], that is, the violation of the Nyquist criterion, at least in parts of k-space, so that the images are reconstructed using a number of samples lower than the one theoretically required to obtain a fully sampled image [143, 6].[1]

In this way, motion artifacts can also be reduced [70] and, in specific circumstances, higher spatial resolution can be achieved [96, 130]. However, undersampling implies other image artifacts, often in the form of aliasing, streak structures, or blurring.

How undersampling can be implemented? The measurement time of a single trajectory can be made short, but before starting to sample a new trajectory, it is necessary to wait for the nuclear spins to recover toward thermal equilibrium. Therefore, the only way to speed up acquisition, by implementing undersampling, is to reduce the overall waiting time by using fewer trajectories, possibly with increased curvature to ensure each trajectory would individually cover a larger portion of the k-space.

As we have described earlier, collecting k-space data involves following sampling trajectories defined by the temporal modulation of gradients.

Although trajectory omission achieves the primary goal, that is, more rapid measurements, it entails violation of the Nyquist criterion, giving rise to two problems. The first is the selection of the optimal scanning scheme in the k-space, that is, the problem of finding the sampling trajectories that cover k-space most densely using fewer trajectories. The second is the problem of omitting as many trajectories as possible without reducing image quality.

Three alternative shapes of sampling trajectories have dominated the literature and are used in commercial scanners, namely, cartesian, radial, and spiral (see Figure 3.8). The issues associated with reducing the number of trajectories and increasing the coverage of each trajectory differ somewhat between these strategies. Whereas cartesian and radial (projection) acquisitions follow straight-line trajectories, spirals illustrate that increasing the sampling trajectories curvature is a powerful means to increasing the coverage attained per trajectory and hence of compensating for the reduced number of trajectories being sampled.

Cartesian trajectories traverse k-space from edge to edge on parallel lines and samples the k-space center just once. An obvious advantage of Cartesian scanning is that all positions coincide with a uniform rectangular grid,

---

[1] Parallel imaging methods such as sensitivity encoding (SENSE) [111], simultaneous acquisition with spatial harmonics (SMASH) [129], and generalized autocalibrating partially parallel acquisition (GRAPPA) [38] also can be thought of as undersampling methods in which artifacts are removed by using the information obtained from multiple RF coils. But, at the end of the process, a complete image is reconstructed by using the partial information collected by different receivers.

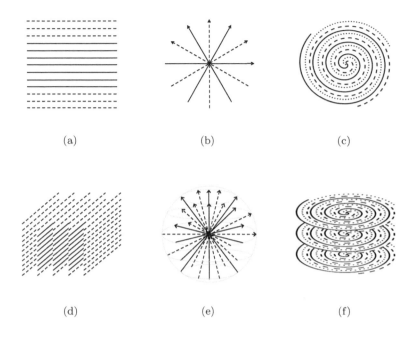

Figure 3.8

Sparse sampling strategies in 2D (a–c) and 3D (d–f) settings: Cartesian (a,d), Radial (b,e) and Spiral (c,f). Measured trajectories are represented by continuous lines; missing trajectories are represented by dashed/dotted lines.

allowing very fast image reconstruction by a simple FT. Acquisition errors, due, for example, to motion, have little effect if they occur during the sampling of peripheral areas of k-space, where the signal is weak, but can produce serious artifacts if they occur during the sampling of the k-space center. Moreover, sampling low and high frequencies (where the signal is very weak and is comparable with noise level), both with the same density, greatly reduces image SNR. Strategies for reducing the number of trajectories include omission of extreme lines asymmetrically, that is, half of k-space (half scan) [47], or symmetrically on both halves (partial Fourier) [79]. Prolonging the trajectory while maintaining the cartesian sampling can be achieved by adding additional phase-encoding steps and refocussing (gradient and/or spin echoes) to obtain multiple lines (fast spin echo imaging, FSE [46], or gradient and spin echo imaging, GRASE [93]) or all lines (echo-planar imaging, EPI [78]) per excitation performed. In the sense of curving the trajectory, these echo trains are achieved by introducing very sharp curvature at the edges of the sampled k-space, with the benefit of retaining the simplicity of reconstruction from a cartesian grid.

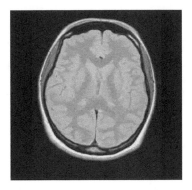

Figure 3.9
A completely sampled (256 × 192 samples) MRI image of the brain of an healthy volunteer collected by a GE Signa MRI System at 1.5 T. For convenience, it has been zero-filled (with 32 columns of zeros to the left and 32 columns of zeros to the right) to obtain a 256 × 256 image.

Interesting radial trajectories start at the center of k-space where the signal is strongest. As discussed in Chapter 2, the strongest advantage of radial scanning are the diminished sensitivity to patient movement. A disadvantage of radial scanning is that data are not amenable to a direct application of the FT, which significantly slows the reconstruction process. Another point is that achieving Nyquist in all parts of the k-space requires lots of projections (more than for cartesian). Obviously, this would increase acquisition time but heavily over samples center of k-space, resulting in greater SNR and lower sensitivity to motion (by averaging it out). Reducing density at center of k-space to meet Nyquist there leads to undersampling of the periphery but can be much faster than cartesian, and the resulting artifacts are reduced.

Finally, spiral trajectories, being strongly curved, cover much more k-space than cartesian and radial trajectories. As a result, fewer spirals are needed to satisfy the Nyquist sampling criterion throughout the k-space. Moreover, as with radial scanning, movement effects and noise occurring at k-space center can be reduced because the center is oversampled: noise is averaged and the resulting image has a SNR greater than a cartesian sampled image.

Figure 3.9 shows a completely sampled (256 × 192 samples) MRI image [110] of the brain of a healthy volunteer collected by a GE Signa MRI system at 1.5 T. For convenience, it has been zero-filled (with 32 columns of zeros to the left and 32 columns of zeros to the right) to obtain a 256 × 256 image to be used as a test image. The zero-filled columns appeared as black bands on each side of the image. This occurs because the original image background had positive value, representing the average noise level of an image used in magnitude form.

Based on this image, we can show the effect of undersampling on different k-space sampling strategies. Starting from this magnitude image, we performed a FT to obtain its k-space samples.

Direct use of these Fourier coefficients would not reflect real-world conditions because the coefficients are obtained from the FT of a real-values object (the magnitude image). These coefficients would exhibit Hermitian symmetry (see Chapter 1) that could affect the performance of the algorithms. In experimental conditions, however, the Hermitian assumption is violated due to resonant frequency variations arising from thermal instabilities as well as artifacts from physiological motion and flow. As a result, phase errors are always present on experimental data.[2]

In order to approximate experimental conditions, after the FT of the given image we introduced random phase errors to prevent the Hermitian symmetry; in this condition, data coming from the four k-space quadrants must therefore be sampled.[3] These data were used to extract samples for cartesian, radial and spiral undersampling. Images obtained by using undersampling (15,360 samples of the 65,536 in the original) are reported in Figure 3.10. To better highlight artifacts, images obtained as differences between the fully sampled reference image of Figure 3.9 and those in Figure 3.10 are shown in Figure 3.11. As can be observed, the three sampling schemes produce different effects: cartesian sampling produces blurring in the direction of the missed trajectories; projection sampling produces mainly "star" artifacts; spiral sampling produces generalized blurring.[4]

The choice of projection or spiral sampling can also be favored by the nature of the observed sample or by sampling time reduction requirements. For instance, in presence of very short T_2^* there is no time to go to the edge of k-space before start a Cartesian trajectory, forcing one to use rapid radial acquisition of FIDs. Spiral scans, whether using constant angular velocity

---

[2]Phase errors can be estimated and corrected if Hermitian assumption can be done. As an example, if image phase is assumed to be smoothly varying over space, the standard approach to phase correction involves the measurement of a symmetric low-frequency k-space portion and the estimation of the image space phase solely from this restricted measurement set. The image is then derived from an undersampled subset of k-space, which also includes the small support used in the phase estimation step. A complex image is then formed by conjoining the image magnitude and phase estimates, with just the real portion of this generated image being retained as the solution. Due to the necessity of acquiring a symmetric, low-frequency spectral band necessary in the phase estimation process, methods relying on Hermitian symmetry such as POCS (projection onto convex set) and homodyne detection [21, 41, 91] can decrease the number of required measurements by less than half of that delimited by Shannon's theorem.

[3]This operation will have been carried on for all MRI images used as samples in this book unless stated otherwise.

[4]Here we are interested only to a qualitative effect, not to the amount of residual artifacts. For the three acquisition schemes, the number of collected samples is insufficient for a correct reconstruction and differences are not due to the number of samples but on the k-space positions where the different trajectories collect the samples.

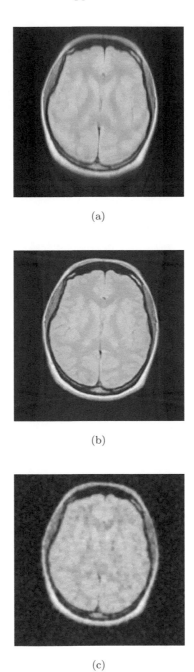

(a)

(b)

(c)

Figure 3.10
Images obtained with (a) cartesian, (b) radial, and (c) spiral trajectories undersampled to use just 15,360 samples, as opposed to 65,536 samples in Figure 3.9.

(a)

(b)

(c)

Figure 3.11
Images obtained as difference between that shown in Figure 3.9 and those in
Figure 3.10(a), Figure 3.10(b) and Figure 3.10(c). The results are shown from
(a) to (c), respectively.

or constant linear velocity, represent as good a trade off between time and artifacts reduction, as they have low motion sensitivity similar to projection sampling, but allow more uniform k-space covering, similar to Cartesian grid sampling.

In MRI applications requiring real-time imaging that would suffer from physiological movements (e.g., cardiac or functional brain imaging), the trade-off between the reduction of movement artifacts by reducing acquisition time, and the avoidance of artifacts due to sparse sampling, becomes particularly important. The main result of the omission of scan trajectories is that there are fewer samples in k-space than the number needed to estimate all pixel intensities in image space. Therefore, an infinity of possible images satisfy the given sparse data and the reconstruction problem becomes ill-conditioned. Additionally, omissions often cause violation of the Nyquist sampling condition, thus producing aliasing. An exciting area of research at present is that of tailoring acquisition to provide the most informative samples in the least time. Chapter 5 will describe some unconventional strategies to minimize the artifacts, which may arise due to use of these strategies.

3.4 Summary

MRI artifacts occur because one or more of the imaging principles have been violated. For conventional imaging techniques, three sources of artifacts are common: frequency displacement due to residual static magnetic field inhomogeneity, motion, and undersampling. Understanding the causes of MRI artifacts will lead to better acquisition sequences designs and the selection of the appropriate correction method.

Artifacts shapes are strongly dependent on the k-space trajectory used. When their effect is mild, they can be efficiently reduced through hardware calibration, imaging parameters optimization, or by using post-processing methods. When the residual inhomogeneity of the main magnetic field is high (tens of ppm or higher), it can, however, be impossible to correct them through post-processing methods or simple modification of imaging parameters. In these cases, completely new (unconventional) imaging strategies and sequences have to be adopted, as will be discussed in Chapter 4.

When sample motion is rapid or dynamic processes are to be imaged, completely sampled trajectories cannot be used and undersampling is necessary. Some innovative imaging strategies to avoid motion/undersampling artifacts are considered in Chapter 5.

Part III

Advanced Solutions

4

Methods for Magnetic Field Inhomogeneity Reduction

CONTENTS

As described in Chapter 3, magnets are often narrow and uncomfortable to ensure high homogeneity. Despite this, residual static magnetic field inhomogeneities continue to be present and are causes of image distortions and artifacts.

Aim of this chapter is to describe an innovative method to reduce artifacts due to residual static magnetic field inhomogeneity, starting by a rapid review of other previously proposed methods on the same subject. These other methods range over a wide variety of subjects (field map, field gradients modulation, amplitude-modulated field gradients pulses), and their brief presentation serves as an introduction.

The proposed unconventional coding/reconstruction algorithm, allowing to the class of amplitude-modulated field gradients pulses methods and based on the assignment of different time-varying frequencies (accelerations) to different spatial positions, is described in detail. The technique can be used both for coding and decoding the signal.

Numerical simulations of the 1D case, also in presence of noise, are reported and compared with those obtained by conventional coding/decoding methods to demonstrate its applicability and efficacy. Nevertheless, this method is treated in a speculative sense; it has not yet been experimentally demonstrated.

A detailed description of the conventional methods can be found in the cited references.

4.1   Introduction

MRI has seen steady progress regarding magnets, acquisition sequences, electronics, software, and, in turn, new applications have opened to the technique. But the initial acquisition paradigm has remained the same; it requires patient insertion in a, as much as possible, homogeneous and stationary magnetic field to which linear magnetic field spatial variations (gradients) are added to encode the signal by assigning different frequencies to different spatial districts.

The amplitude of the signal at each frequency is proportional to the spin density present in the sample region of corresponding magnetic field. The collected signal represents the envelope of a series of signals occurring at different frequencies, from different spatial districts. The advantage of this method is its simplicity; an FT of the measured signal suffices to provide the spatial distribution of spins if the field varies linearly with position.

The signal is extremely sensitive to residual static magnetic field inhomogeneity, resulting in frequency displacement that corresponds in distortion of the reconstructed images, that is, spatial displacement of pixels from their theoretical positions. The decoding system does not eliminate the effect of signal displacement if residual, static magnetic field inhomogeneities are present.

To reduce this unwanted effect as much as possible, magnets have to be constructed with very high magnetic field homogeneity in the FOV and shimming applied. For this reason, magnet construction is a very long and expensive process requiring:

1. Accurate design;

2. Precise mechanical construction (due to the unavoidable mechanical construction tolerances and to impurities in the materials used, the residual magnetic field inhomogeneity can remain too high);

3. The use of a shimming procedure.

Another way to obtain a more homogeneous magnetic field inside the useful FOV could be to increase the whole magnet volume, but such a solution should be even more expensive.

In what follows, a description of methods that can be used to overcome inhomogeneity artifacts is given. In particular, the methods are grouped in two classes: the first class is comprised of methods used in conventional, frequency encoding, imaging; the second considers just one innovative method based on temporal frequency coding and an appropriate decoding scheme, which has not yet been experimentally demonstrated. For this reason, much of this chapter is reserved to this speculative method.

Most of the reported methods are protected by patents and, for this reason, most of the references on these arguments are constituted by patents. Part of the following material has been reproduced from [109] (with kind permission from Bentham Publisher: *Recent Patents on Biomedical Engineering*, Recent patents on magnetic resonance imaging sequences in presence of static magnetic field in-homogeneity, 2(1), 2009, 73–80, G. Placidi, D. Franchi, A. Maurizi, and A. Sotgiu).

4.2 Conventional Methods

4.2.1 Field Mapping

Methods based on magnetic field mapping were the first approaches to correcting for inhomogeneity effects to be presented in literature; once the nature of the artifacts has been characterized, numerical techniques to estimate and compensate these effects can be applied directly to the reconstructed, distorted, images.

Yao et al. [154] and Kaufman et al. [62] described the use of calibration data derived from an extra measurement cycle, taken without the application of field gradients, to compensate for static magnetic field drifting during data measuring process. Beside the correction of frequency shift, these data were also used for correcting phase shifts.

A subsequent Kaufman–Carlson [61] method dealt with magnetic field inhomogeneities in form of constant magnetic field gradients along each spatial direction. The presence of such static gradients will affect the collected images especially along the phase-encoding directions. Their method consisted of determining the displacement of the maximum amplitude of the MRI echo from the zero-phase-gradient position along the phase direction and compensating for this error in a successive data acquisition. The method can be applied directly to the data used to reconstruct a given image (no additional data are necessary), but it is able to correct only distortions produced by inhomogeneities affecting the phase-encoded directions and generated by constant magnetic field gradients.

The method presented by Meyer et al. [85] is one of the most effective for the correction of linear magnetic field inhomogeneity. It includes the calculation of a local field map (or distribution), finding the best linear fit, and using this linear field map to correct image distortions due to local frequency variations, through mathematical deblurring.

The MR signal emitted by an object $m(x, y)$ in a field with local frequency deviations from the demodulation frequency is

$$s(t) = \int \int m(x, y) e^{-i2\pi(k_x x + k_y y + t f(x, y))} dx dy \qquad (4.1)$$

where $f(x, y)$ is the field distribution, k_x and k_y are defined as the area under the readout gradients.

The local field map $f(x, y)$ in the presence of field inhomogeneity is calculated from two images collected at different echo times. If the first image is

$$M_1(x, y) = m_1(x, y)e^{i\phi_1(x,y)} \qquad (4.2)$$

and the second image, collected at an echo time Δt time later, is

$$M_2(x, y) = m_2(x, y)e^{i\phi_2(x,y)} \qquad (4.3)$$

the field distribution is given by

$$f(x, y) = (\phi_2(x, y) - \phi_1(x, y))/(2\pi\Delta t). \qquad (4.4)$$

It is difficult to reconstruct the image $m(x, y)$ from Equation 4.1 for arbitrary field inhomogeneity $f(x, y)$. For this reason, the solution of a linear approximation is commonly used. The field distribution can be obtained by echo-planar or by spin echo imaging. The linear fitting field map is then

$$\tilde{f}(x, y) = f_0 + \alpha x + \beta y. \qquad (4.5)$$

The linear field distribution, from the field measurements, is determined using a maximum likelihood estimator [116] with weights proportional to the pixel intensity. Linear assumption for $f(x, y)$ represents a simplification allowing to correct only particular cases of field inhomogeneity.

By applying linear assumption, the received signal from an object in presence of linear field variation described by Equation 4.5 can be expressed as

$$s'(t) = e^{-i2\pi t f_0} \int \int m(x, y)e^{-i2\pi(k'_x x + k'_y y)} dx dy \qquad (4.6)$$

with $k'_x = k_x + t\alpha$ and $k'_y = k_y + t\beta$. According to the model, data are reconstructed assuming there is a linear field variation that can be compensated by using the new trajectories (k'_x, k'_y) and demodulating to a frequency of f_0 in Equation 4.6. It is important to note the method imposes a heavy computational overhead and is not effective in treating low SNR regions and abrupt field changes.

Harvey [45] proposed a method based on the collection of two images by using a single acquisition sequence composed by a series of readout gradient pulses of opposite sign. Positive pulses are used for one image; negative pulses are used for the other. At each cycle of the readout gradient, a phase-encoding pulse is applied to change row in the k-space. Each line of the k-space is covered both from the left and from the right. The time difference between the two images, I_1 and I_2, is a fixed and known delay time Δt.

Another difference between the two images is the phase dispersion, which differs from pixel to pixel. Both phase difference and time delay are used, point by point, to determine the local magnetic field inhomogeneity map by using the equation:

$$\gamma\Delta B(x, y, Z_0) = arg(I_2(x, y)/I_1(x, y)). \qquad (4.7)$$

The estimated inhomogeneity map is used to correct image distortions mathematically (it can also be used to determine currents for shimming coils, but that is out of the scope of this book).

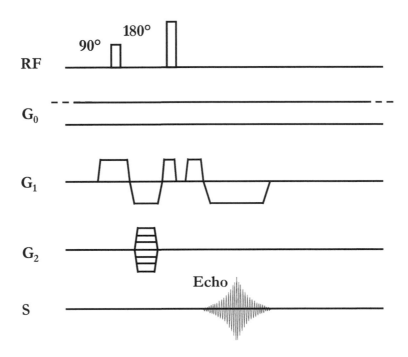

Figure 4.1
The imaging sequence introduced in [33] to deal with magnetic field inhomogeneity.

Epstein et al. [31, 32, 33] proposed methods to deal with special types of magnetic field $\mathbf{B}_0$ inhomogeneities: those in which there is a single critical point within the FOV (the field deviation has a minimum or a maximum within the FOV) and those in which the magnetic field is characterized by a monotonic field gradient $\mathbf{G}_0$ along the main field direction. The first is treated by selectively exciting only spins lying in a small neighborhood of a local point; in this way, one can do spatially localized, high SNR, spectroscopic (single point imaging) measurements without the need for further spatial encoding. The monotonic gradient on the other hand is dealt with using a sequence whose scheme is reported in Figure 4.1. Here, $\mathbf{G}_0$, the permanent gradient due to residual magnetic field inhomogeneity, is used to codify spatial information when combined with other adjustable gradients, $\mathbf{G}_1$ and $\mathbf{G}_2$.

It can be assumed that z the direction of $\mathbf{G}_0$ and x and y the directions of $\mathbf{G}_1$ and $\mathbf{G}_2$, respectively. The sequence is essentially a 2D spin-echo sequence in which, during slice selection and the refocusing pulse, the combined gradient $\mathbf{G}_0 + \mathbf{G}_1$ is used to select the slice. In the readout interval, a gradient $\mathbf{G}_0 - \mathbf{G}_1$ is used for frequency encoding. Phase encoding involves the combination $\mathbf{G}_0 + \mathbf{G}_2$. The used gradients have the effect of slanting the slice with respect to its theoretical orientation (if the residual inhomogeneity was zero, the slice would be parallel to the $x - y$ plane). Figure 4.2 shows this situation. In particular, $\mathbf{G}_0 + \mathbf{G}_1$ implies a rotated slice. $\mathbf{G}_0 - \mathbf{G}_1$, the readout direction, is not necessarily orthogonal to slice (Figure 4.2(b)), unless $|\mathbf{G}_1| = |\mathbf{G}_0|$ (Figure 4.2(a)). Similar consideration applies to the phase-encoding gradient $\mathbf{G}_0 + \mathbf{G}_2$. The line of integration, defining the content in each pixel of the slice, is not necessarily parallel to the slice selection gradient (Figure 4.2(b)); this can reduce the resolution along the readout direction unless the slice is not sufficiently thin. Since the encoded signal may be interpreted as samples of the Fourier transform of a function of three physical variables, an image can be recovered by using standard Fourier inversion techniques. The gradients lie on the same plane, that defined by $\mathbf{G}_0$ and $\mathbf{G}_1$, but have different directions.

In principle, there is no difficulty in obtaining useful imaging measurements with this approach, even in the presence of a permanent gradient along the direction of $\mathbf{B}_0$. In practice, the method has some drawbacks; it allows the elimination of artifacts only in a particular case of magnetic field inhomogeneity; it performs well when the residual magnetic field inhomogeneity strength is no more than six times greater than the maximum strength of the adjustable gradients [32]; it is forced to maintain a thin slice thickness to ensure the desired readout resolution when the residual inhomogeneity strength is higher than the maximum strength of the adjustable gradients.

The field mapping approaches also include specific methods for echo-planar imaging proposed by Reese et al. [116] or by Jezzard et al. [57, 149]. Though very accurate and fast, these methods are designed for a specific acquisition strategy, echo-planar imaging ([116, 57, 149]), or devoted to a specific application, functional MRI of the human brain ([57, 149]). Further details can be found in the cited references.

4.2.2   Field Gradients Modulation

More interesting methods use modulated field gradients; they perform volumetric imaging from a single signal, collected in a short time, by using localization functions.

Some methods [9, 156, 24, 25] consider the problem described by Epstein [32], using, in different forms, pulses of gradients for spatial encoding and re-

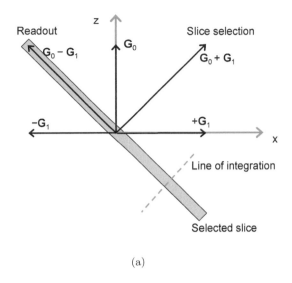

(a)

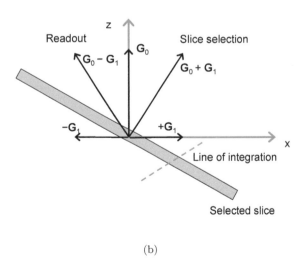

(b)

Figure 4.2
Slant slice imaging [32] due to residual inhomogeneity (a) with adjustable gradients of equal strength to the permanent gradient and (b) with adjustable gradients of smaller strength than the permanent gradient. In these figures, $\mathbf{G}_y$ is considered to be orthogonal to the xz-plane. Line of integration has also been indicated.

focusing pulses to repeatedly refocus the accumulating phase dispersion along the direction of a permanent gradient. In these cases, less homogeneous fields are treatable.

Noll et al. [92] described a method that reduces blur due to off-resonance reconstruction through the reconstruction of a series of basic images after the demodulation of the signal at different frequencies. In each of these basic images, a focusing measure is calculated to determine which regions are in focus and which are blurred. Finally, an unblurred composite image is created by combining the unblurred regions of the basic images. The method is sketched in Figure 4.3.

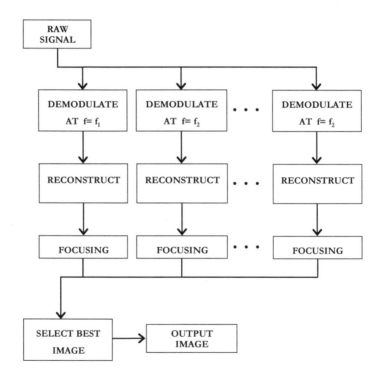

Figure 4.3
Demodulation/reconstruction process described in [92].

The processing algorithm is completely automatic and does not require a map of the local resonant frequencies. Limitations of the algorithm are the computational overhead; the limited degree of inhomogeneity it eliminates (650 Hz maximum, in a FOV of 45 cm) and the fact it reduces just blurring. This method is unable to reduce higher order or different shape of magnetic field inhomogeneities that produce spatial distortions or image folding.

Yang et al. [153] proposed a method for processing gradient echo data collected for individual slices. It serves to retain image portions masked by intra voxel phase dispersion mainly produced by localized differences in magnetic permeability/susceptibility.

The method consists of the application of an additional phase-encoding gradient after the slice selection; data corresponding to the same slice are acquired for different steps of the additional phase-encoding gradient. It is possible to recover a series of images, for the same slice, by performing a third Fourier transform along the slice selection direction. The image corresponding to the zero value of the additional gradient contains the complete signals of those voxels that have not been affected by local magnetic permeability/susceptibility; the other images contain signals for voxels subject to local phase dispersion due to permeability/susceptibility differences.

The corrected phase dispersion image is the sum, pixel by pixel, of the set of images collected for different phase gradient values applied along the slice selection direction. The method assumes a slice selection gradient is applied along the direction of the main magnetic field (the selected slice has to be perpendicular to the field direction) and makes use of a gradient, to codify the slice phase dispersion, necessary to be applied along the slice selection direction (the same of the main magnetic field). The method is very effective in reducing permeability/susceptibility artifacts.

However, the SNR can be lowered and image contrast reduced in areas where local permeability differs markedly, such as adjacent to the peripheries of organs, cranial sinuses, and similar locations. Moreover, it reduces phase dispersions just along the main magnetic field direction.

Janzen et al. [56] proposed an articulated procedure to cope with a series of magnetic field inhomogeneity. It is summarized in Figure 4.4. In particular, the method treats the following three situations:

1. Static magnetic field inhomogeneity (indicated by A in Figure 4.4)

2. Inhomogeneity in the gradient field (mainly affecting the slice thickness, indicated by B in Figure 4.4)

3. Magnetic field susceptibility differences which produce local dephasing and, hence, MRI signal losses (indicated by C in Figure 4.4)

The correction procedure consists on a series of decisions about the presence of inhomogeneity of type A, B, C, or their combinations and the applications of the appropriate corrections. Correction A consists of the use of a compensation gradient along the slice selection direction during the application of the readout gradient (having the same amplitude as the slice selection gradient).

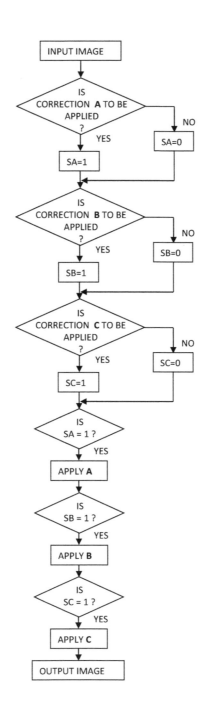

Figure 4.4
Magnetic field inhomogeneity correction method used proposed by Janzen et al. [56].

It is supposed the slice selection gradient is applied along the main magnetic field.

Correction B consists on the acquisition of two images, for the same slice, using identical pulse sequences except the slice selection gradient sign. Only the gradient inhomogeneity, occurring along the slice selection gradient, is considered. The correction for B is applied by considering the two collected images, I_1 and I_2, to form the final image I as follows:

$$I = 2I_1 I_2 / (I_1 + I_2). \tag{4.8}$$

This is because when a gradient error ΔG is added in the slice selection gradient, suppose $\mathbf{G}_z$, the slice thickness becomes $d = \Delta \omega / \gamma (G_z + \Delta G)$ instead of the nominal $d = \Delta \omega / \gamma G_z$, where $\Delta \omega$ is the bandwidth. This implies a change in the slice thickness. The acquisition of two images at opposite slice selection gradient values yields two images of slice thickness $d_1 = \Delta \omega / \gamma (G_z + \Delta G)$ and $d_2 = \Delta \omega / \gamma (G_z - \Delta G)$, respectively. The desired nominal slice thickness $d = \Delta \omega / \gamma G_z$ can be obtained if $d = 2d_1 d_2 / (d_1 + d_2)$. Because the image intensity is proportional to the slice thickness, the corrected image is given by $I = 2I_1 I_2 / (I_1 + I_2)$.

Correction C consists of the acquisition of two images with the addition of positive and negative offset slice selection gradient lobes, respectively, applied just before the readout interval (by regulating duration and amplitude so that each lobe gives a 50% signal reduction in a region of uniform magnetic field). The same assumptions of case B apply.

Correction C is applied from the two collected images, I_1 and I_2, forming a final image $I = I_1 + I_2$. In fact, in a region of signal loss due to dephasing, the image obtained by using positive gradient offset lobe is brighter than normal, while the image obtained using negative gradient offset lobe is darker than normal. The two images, though both abnormal, are complementary to each other. Thus, a simple addition of these two images yields an image in which the intensity is uniform across the entire FOV.

The main drawback of this method is that it is very difficult to evaluate the presence of the three forms of inhomogeneity. For this reason, the application of all the corrections is generally required, thus increasing acquisition time. Moreover, it does not have a unified strategy to treat magnetic inhomogeneity. Lastly, the method relies on some restrictive assumptions about the gradients application and inhomogeneity distribution.

4.2.3 Some Remarks

Though the previous methods are very accurate in correcting some types of inhomogeneity artifacts, they have the following limitations:

1. They are usable only for particular inhomogeneity.

2. The inhomogeneity must assume specific directions (e.g., to allow

phase-encoding gradients to be used along the inhomogeneity directions).

3. They are applicable only to sequences that use phase-encoding gradients.

4. They treat only relatively small inhomogeneities problems (of the order of 1 ppm that, for a MRI equipment operating at 1.5 T, corresponds to 15 mG, about 100 Hz).

Thus, an unsolved problem remains: how to perform imaging in the presence of relatively high residual static magnetic field inhomogeneity.

4.3 An Unconventional Solution

In what follows, we discuss an innovative paradigm for MRI by investigating acquisition, coding, and reconstruction strategies for imaging unconstrained by static magnetic field residual inhomogeneity.

The proposed technique, based on references [99, 107, 108] (some of the material reported therein has been reproduced with kind permission from Springer Science+Business Media: Proceedings of the 4th International Symposium on Advances in Visual Computing, Part II, A novel acceleration coding/reconstruction algorithm for magnetic resonance imaging in presence of static magnetic field in-homogeneities, ISVC '08, 2008, 1115–1124, G. Placidi, D. Franchi, A. Galante, and A. Sotgiu), substitutes the usual signal frequency coding scheme with one based on temporal frequency variations, through the application of temporally variable gradient pulses.

The main idea of the method is that a measurement of the frequency change rate when applying time-varying gradients allows us to unambiguously assign the signal amplitude/phase to a given spatial source. The method is different from others using modulated gradients as:

1. It does not use a frequency coding scheme;

2. It does not use a single signal per image;

3. It does not use localization functions to obtain reconstruction;

4. It does not use long duration gradients, but extremely rapid, amplitude modulated, gradient pulses;

5. It does not need repetition of the imaging acquisition process to estimate residual magnetic field inhomogeneity;

6. It does not depend on the magnetic field inhomogeneity distribution, shape, and intensity.

In principle, it should allow imaging from completely open flat magnets that produce highly inhomogeneous static magnetic field, whose VOI is not required to be confined inside the reduced space between poles or at the center of a cylindrical magnet.

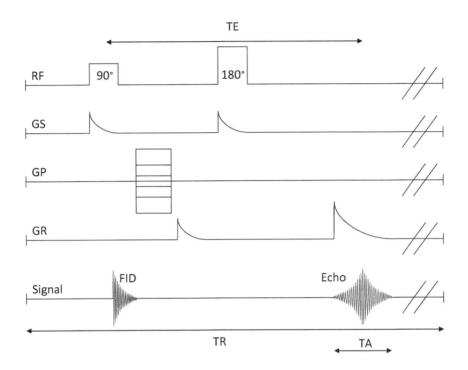

Figure 4.5
Example of an acceleration acquisition sequence. Note the shapes of the gradients (the phase-gradient steps reproduce the same nonlinear shape).

4.3.1 Spatial Encoding by Nonconstant Gradients

We can start with a problem definition and model. As the model, we consider the MRI signal of a sample when using conventional imaging. The spatial frequency variation, in the 1D case suppose along x, is

$$\omega(x) = \omega_0 + \omega_{res}(x) + \gamma Gx \qquad (4.9)$$

where x represents the encoding direction, G the gradient amplitude along x, γ the gyromagnetic ratio, ω_0 the precession frequency corresponding to the main magnetic field $\mathbf{B_0}$, and ω_{res} the residual frequency due to the magnetic field inhomogeneity produced at point x. Here we assume the gradient value is unchanged during signal acquisition. As described above, signal is sensitive to magnetic field inhomogeneity that can lead to distortions in the reconstructed images.

After demodulation with ω_0, the signal coming from the spatial position x is

$$s_x(t) = Ae^{-t/T}e^{-i(\omega_{res}(x)+\gamma Gx)t} \tag{4.10}$$

where T is the relaxation time (depending on the used sequence, we can have T_2 or T_2^* and, for this reason, we indicate it generically with T) and A the signal amplitude (proportional to the spin density) at the position x. The term ω_{res} gives a frequency displacement, and this leads to a spatial displacement of the pixel x from its theoretical position. The coding system does not eliminate the effect of signal spatial displacement if magnetic field inhomogeneity is present.

The proposed method substitutes the usual constant frequency encoding with one based on frequency temporal variations, through the application of gradient pulses which are temporally variable (see, for example, Figure 4.5). In this case, Equation 4.9 becomes

$$\omega(x,t) = \omega_0 + \omega_{res}(x) + \gamma G(t)x. \tag{4.11}$$

The corresponding demodulated resonance signal will be

$$s_x(t) = Ae^{-t/T}e^{-i(\omega_{res}(x)+\gamma G(t)x)t}. \tag{4.12}$$

The inhomogeneity contribution is still present, but it differs from the contribution due to the gradients in that the former is time independent, while the latter has a temporal variation. Thus, the constant term can be eliminated if an appropriate signal analysis method is used.

One choice for the gradient is

$$\gamma G(t) = at^{(1-k)/k} \tag{4.13}$$

is the temporally varying gradient intensity (with a and k constants, a includes also the γ contribution and $k \geq 2$) applied along x.

It is clear from Equation 4.12 and Equation 4.13, that the resulting spin signal from a position x has a time-varying frequency (a chirp) whose shape depends on the gradient shape defined in Equation 4.13. The starting frequency depends in an unknown way on the local field (i.e., it depends on the $\omega_{res}(x)$) whilst the frequency change rate has a well established dependence on position (i.e., ax).

If the signal is expressed as a function of a new time scale $\tau = t^{1/k}$, we get

$$s_x(\tau) = Ae^{-\tau^k/T}e^{-i\omega_{res}(x)\tau^k}e^{-iax\tau}. \tag{4.14}$$

If the acquisition interval is small enough to consider $\omega_{res}(x)\tau^k$ constant, the signal amplitude of each spatial point is encoded at a different frequency ax that does not depend on the residual magnetic field inhomogeneity.

For the used encoding method, we use the term "acceleration encoding" to indicate that temporal variation of frequency is used to encode the signal instead of constant frequency. In fact, in conventional sampling of the signal, the fixed strength gradient traverses equal distances in k-space in equal times. In the phase-encoding direction, the application time is fixed and the gradient stepped by a fixed increment. A gradient that is constant for a time corresponds to a fixed velocity in k-space. When we make the transition to time-varying gradients (Figure 4.5), the analogy is to accelerate the change of position in k-space (hence, the term "acceleration" being applied to such strategy). For phase-encoding, the gradient step need to be ample dependent to replicate the behavior in this direction (see the GP shape in Figure 4.5). The time-varying gradient can be used for slice selection, in which case the transmitted RF pulse can also be designed as a chirp, according to the gradient shape function, by adopting the following function (see Figure 4.6):

$$acc_s(t) = \frac{\sin\left(ht^{(1-k)/k}\right)}{ht^{(1-k)/k}}. \tag{4.15}$$

where h is a constant that determines the bandwidth and k is the same used in the gradient shape, as in Equation 4.13. The function acc_s, an acceleration *sinc*, serves to ensure a slice with sharp profiles in the acceleration domain and has the same role of the *sinc* in the frequency domain (see Chapter 1), that is, to define a deep slice profile.

It remains to consider how this encoding can be used for reconstruction. As illustrated in Figure 4.7, the sampling and reconstruction method consists of the following steps:

1. Measure the signal emitted by the spins (a chirp) at a constant sampling rate for a time interval which is shorter than half the period of the maximum residual inhomogeneity frequency (Figure 4.7, step A). This fulfils the condition discussed above that $\omega_{res}(x)\tau^k$ be constant.

2. The signal is recognized being collected on a nonuniform grid and its points are mapped accordingly from the temporal axis to a gradient moment axis (Figure 4.7, step B).

3. The resulting signal is interpolated to yield a uniformly sampled signal (Figure 4.7, step C); in this way, the chirp assumes the form of a sum of constant frequency waves.

4. Points at a fixed interval are selected (decimation is applied) from the interpolated signal (Figure 4.7 , step D).

The resulting signal is analyzed with FT. Each frequency corresponds to a given acceleration, that is, to a specific spatial position. It is important to

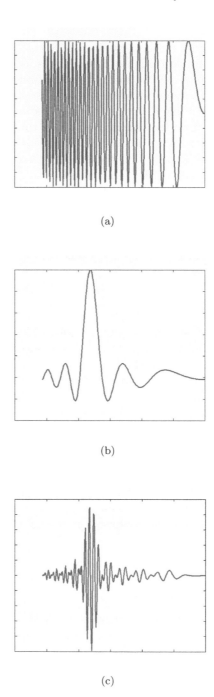

(a)

(b)

(c)

Figure 4.6
RF signal used in coding by acceleration: RF chirp (a); acceleration sinc (b); modulated RF signal (c) obtained by multiplying the chirp (a) with acc_s (b).

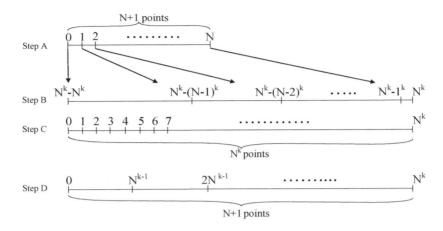

Figure 4.7
Decoding method used to analyze acceleration-coded signals.

note that steps C and D can also be merged into a unique step consisting of interpolating only the desired points as represented in Figure 4.7, step D.

The nonlinear sampling scheme (step B) depends on the gradient shape (in the Figure 4.5, the reported gradient shape corresponds to $k = 2$ in Equation 4.13).

4.3.1.1 Numerical Simulations

Using the MRI simulator described in Chapter 2, the proposed method has been tested and compared with the standard one in presence of a defined magnetic field inhomogeneity condition.

We considered a 1D sample (2D or 3D extensions do not modify the nature of the experiment) composed of three separated small objects P1, P2, and P3, having the same $T_1=T_2=1$ s, in the configuration reported in Figure 4.8.

We assumed their positions with respect to the center of the magnet ($x = 0$) to be equal to 1, 2, and 2.5 cm, respectively, and assigned at each object position a residual static magnetic field inhomogeneity of 0.5, –0.5, and 0 Gauss, respectively. In this case, the whole sample would have a frequency bandwidth of 4257.5 Hz, corresponding to a linewidth of 0.235 mGauss. In the absence of magnetic field inhomogeneity, if a spatial resolution of 0.1 cm is required, a magnetic field gradient of 0.0235 mT/m should be used to overcome the intrinsic linewidth of the signal.

Due to the presence of 1 Gauss magnetic field inhomogeneity on a scale distance of 1 cm, the classical coding technique would require a magnetic field gradient greater than 10 mT/m to separate the objects. However, peaks positions would continue to be wrongly represented, though very high magnetic field gradient is used. The result is shown in Figure 4.9(a).

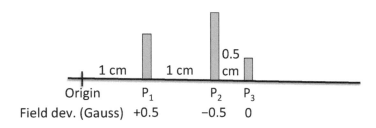

Figure 4.8
Three noise-free peaks, of different spin densities (amplitudes 2, 3, and 1, respectively), positioned along the x-axis; positions and experienced magnetic field deviations, with respect to the central magnetic field value, are also indicated.

For the model object, the traditional acquisition scheme with encoding gradient (10 mT/m) is not sufficient to represent correct peaks positions and separations. This example demonstrates the inefficacy of using a constant gradient for frequency coding in the presence of static magnetic field inhomogeneity.

As previously discussed, to avoid spurious effects using the proposed coding/reconstruction method on the same data, it was necessary to limit the acquisition time to half of the period corresponding to the variation due to the inhomogeneity. As the maximum frequency variation due to inhomogeneity is 4257.5 Hz, corresponding to a period of 230 μs, this can be achieved using a sampling interval of 100 μs.

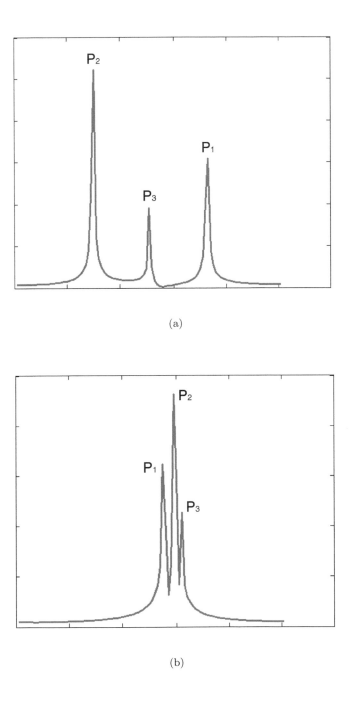

(a)

(b)

Figure 4.9
Spatial reconstructions of the sample shown in Figure 4.8 obtained by signals collected both with conventional frequency acquisition method (a) and with the proposed acceleration acquisition method (b). Reciprocal peaks positions and separations are well reconstructed in (b).

Moreover, a sampling rate of 1 μs (sampling frequency of 1 MHz, 100 points in 100 μs) and a magnetic field gradient of 0.0235 mT/(m*s$^{1/2}$) are chosen. The result obtained, shown in Figure 4.9(b), is very accurate not only for the spatial representation of each object but also for the relative amplitudes. Additionally, a minimal gradient amplitude has been sufficient to separate the objects. The peaks width of the reconstructed signal depend on the used relaxation times.

It is important to note that due to differences in the acquisition parameters only the relative peak positions and heights reported in Figure 4.9(a) and in Figure 4.9(b) can be directly compared.

To complete the analysis, two main points have to be addressed:

1. The choice of the gradients shape

2. The sequence timing

(1) $G(t) = at$, is the simplest gradient shape, but the very rapid variation of the gradient is likely be a problem, especially for severe inhomogeneity, mainly due to the limited amplifiers capacity and to the coils inductance. $G(t) = at^{(1-k)/k}$ might be a good alternative, but it is divergent for $t = 0$, when $k \geq 2$. To solve this problem, it can be treated with a low-pass filter $f(t)$ to obtain

$$G(t) = f(t) \otimes (at^{(1-k)/k}) \tag{4.16}$$

where the symbol $\otimes$ stands for convolution, as defined in Chapter 1. In practice, the maximum bandwidth is fixed to that of the gradient coils/power supplies ensemble and the filter bandwidth used must be greater than the bandwidth of the electronic devices. If these conditions are met, Equation 4.16 represents a smooth function that will tolerated by the gradient hardware.

Other useful gradient shapes are those defined as horizontal flips of the previous one (positive accelerations), as used in the following example.

(2) The sequence timing depends strictly on the T_2^* produced by the field inhomogeneity. For example, supposing a maximum inhomogeneity of 1 Gauss and T_2=1s, $T_2^* \approx 0.81$ ms. In this case, a spatial resolution of 1 mm can be obtained if:

1. The RF pulse duration is about 0.1–0.2 ms;

2. The separation between 90 and 180 pulses is about 0.8 ms;

3. TA = 0.1 ms (sampling frequency of 1 MHz);

4. The gradient value is 0.0235 mT/(m*s$^{1/2}$).

4.3.1.2   Noise Tolerance

To test noise tolerance of the method, numerical noise was added to the sampled data of another numerical phantom (reported in Figure 4.10). Here, the three objects have different amplitudes (9, 4, and 1, respectively). We assumed their positions with respect to the center of the magnet (x = 0) to be equal to 1, 1.5, and 2.5 cm, respectively, and added at each object position a static magnetic field inhomogeneity of 0.5, −0.5, and 0 Gauss, respectively. Again, all the objects had the same relaxation time $T_1 = T_2 = 1$ s and the whole sample would have a frequency bandwidth of 4257.5 Hz corresponding to a line broadening of 0.235 mGauss. As before, standard frequency coding/decoding

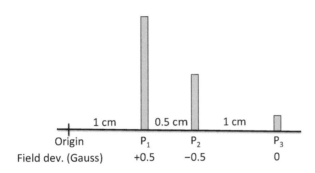

Figure 4.10
Three noise-free peaks, of different spin densities (amplitudes 9, 4, and 1, respectively), positioned along the x-axis; respective positions and experienced magnetic field deviations, with respect to the central magnetic field value, are also indicated.

would produce unavoidable and unacceptable spatial displacement. The simulated acquisition parameters for acceleration imaging were chosen exactly the same as the previous example except that the gradient shape was increasing (positive accelerations have been used) rather then decreasing.

Without loss of generality, white noise was added independently to the sampled real and imaginary signals yielding the signals seen in Figure 4.11. Though present, relaxation time effects are not evident in the signals due to the difference between the high relaxation time and the small sampling period. SNR (defined in Chapter 1) was 10 in each signal channel, a lower bound

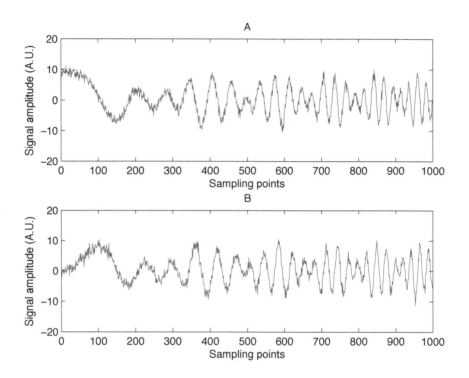

Figure 4.11
Real (A) and imaginary (B) components of a signal collected by the virtual
MRI spectrometer, after the summation of a white noise. Positive accelerations
have been used.

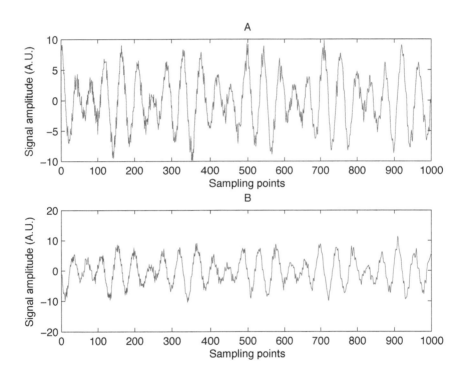

Figure 4.12
Signals of Figure 4.11 after decoding.

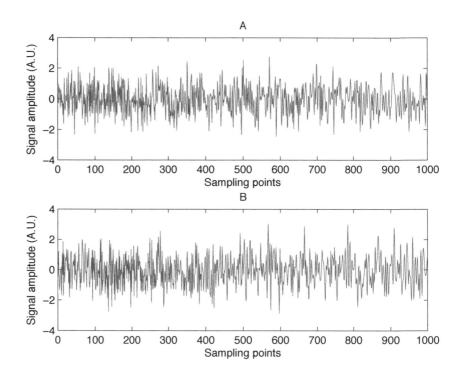

Figure 4.13
Residual noise (real and imaginary parts) obtained by subtracting from the signals reported in Figure 4.12, those obtained through the application of the decoding method to the original MRI signals without the addition of noise.

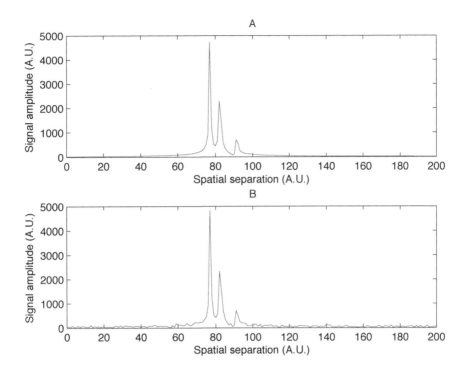

Figure 4.14
Reconstructions of the sample shown in Figure 4.10 by using the proposed method before (A) and after (B) the addition of noise.

in most single shot MRI. The signals resulting from the decoding method (Figure 4.7) are reported in Figure 4.12.

Figure 4.13 shows the signals obtained as a difference between those reported in Figure 4.12 with those obtained after the application of the decoding method on the noise-free signals. The reconstructed spatially resolved signals, before (A) and after (B) noise addition, are shown in Figure 4.14.

Figure 4.15 reports the power spectrum of the original noise (A) and the residual noise (B) after decoding, respectively. Both mean value M and standard deviation σ, were substantially maintained. In fact, the original noise had $M = 0$ and $\sigma = 0.98$, the residual noise had $M = 0$ and $\sigma = 0.91$. Residual noise was reduced especially for high acceleration values, as can be seen by looking at Figure 4.15.

Moreover, residual noise shows a chirp shape (see Figure 4.13) with high frequencies at the beginning and lower frequencies at the end of the readout interval; exactly inverted with respect to the collected signals.

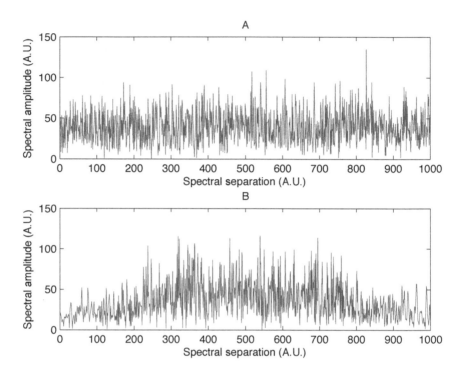

Figure 4.15
Original white noise power spectrum (A) and residual noise power spectrum (B).

This is due to the decoding strategy that operates in inverted mode with respect to the coding gradient; it compresses the signal at the beginning and stretches it at the end of the readout interval (in the case of negative accelerations, opposed situation holds).

Noise has not been increased, moreover the reconstructed spatial signal has not been blurred nor distorted. This demonstrates the coding/decoding process is very precise not only for the spatial position of each object but also for the relative amplitudes, even in presence of noise.

The inhomogeneity amplitude would, however, be responsible for an increased noise in the collected signal (to a greater inhomogeneity would correspond a greater bandwidth of the measured signal, i.e., a lower SNR). After demodulation, the noise level was not amplified by the proposed analysis method. The line broadening in the reconstructed signals depends on the width of the acquisition time interval.

Experimental results are necessary to verify the correctness of the method, to evaluate its limitations, and to optimize the encoding patterns on real scanners for 2D and 3D acquisition sequences.

4.4 Summary

Methods used to reduce artifacts due to residual static magnetic field inhomogeneity have been briefly reviewed. These methods, ranging from field mapping to field gradient modulation and amplitude-modulated field gradients pulses, have been separated into two classes: those used in conventional frequency-encoded imaging sequences and those used in unconventional imaging.

The former methods can be efficiently used when residual magnetic field inhomogeneity is, at most, of the order of 1 ppm and has some particular distributions.

The latter class is based on amplitude-modulated magnetic field pulses. In particular, we considered a specific implementation of frequency temporal variation coded signals, describing both encoding and decoding processes, and testing on numerical 1D data.

The method's tolerance to noise has also been verified. The reported preliminary tests suggest this method could, in principle, be capable to eliminate also residual inhomogeneity of the order of 1 Gauss, which would correspond to 100 ppm at 1 T.

The results are very promising and demonstrate that a high residual magnetic field inhomogeneity can be eliminated without noise amplification, signal distortion, or blurring. Moreover, the technique is independent of residual magnetic field inhomogeneity shape and distribution. Numerical simulations have been carried out with acquisition parameters achievable by many commercial MRI scanners.

In principle, the ability to obtain viable images under high field inhomogeneity could reduce the costs of magnet construction and shimming, allow the construction of more accessible magnets, increase the FOV of existing scanners, reduce chemical shift and magnetic susceptibility effects, and reduce imaging time.

Other methods [8, 94, 123] have been recently proposed to perform very fast imaging by using chirp-modulated RF pulses during slice selection. These techniques are provided to be less sensitive to magnetic field inhomogeneity than those using conventional RF pulses. In common with these methods, the acceleration-encoding method described above uses chirp-modulated RF pulses; the main difference is that the described method uses monotonically varying readout gradient to produce also a chirp-modulated readout signal (to which a demodulation method has to be applied before Fourier reconstruction). The principle of chirp modulation could be the right way to increase MRI tolerance to residual magnetic inhomogeneity.

A lot of practical work remains regarding experimental measurements, to test new acquisition sequences based on this method, to optimize their parameters for commercial MRI scanners, and to perform 2D and 3D experiments.

5

Methods to Handle Undersampling

CONTENTS

In what follows, some interesting methods for sparse sampling (undersampling) acquisition and reconstruction are presented. In particular, a differentiation is made between those reconstruction/restoration methods that modify the k-space trajectories during acquisition, that is, the chosen trajectories (both in number and directions) depend on the sample shape, and those methods that reduce artifacts independently of the sample shape. The first class contains adaptive methods, the second includes compressed sensing. As a further examination, the possibility of jointly using adaptive methods in a compressed sensing strategy has been also explored and some preliminary results, demonstrating that these joint methods can allow a further reduction of the dimension of the data set necessary for good reconstruction, are reported. The implementation and practical impact of sparse sampling methods are also discussed.

5.1   Introduction

As described in Chapter 3, the time required to fully sample k-space a Cartesian grid is relatively long. A common approach is to use undersampling to reduce the number of readout lines in the grid. Alternative non-Cartesian trajectories can provide faster k-space coverage and more efficient gradients usage. When very fast volume coverage is required, undersampling strategies can be combined with non-Cartesian trajectories for further scan time reduction. Undersampling can influence the resulting image in different ways, depending on the details of the k-space covering paths: Cartesian, radial, spiral, 2D, 3D, etc. (see from Figures 3.9 to 3.11). This stems from how undersampling is achieved with the different strategies of k-space sampling. For example, when Cartesian sampling is used, undersampling occurs by missing some rows along the phase-encoding direction in the k-space rectangular grid; in case of radial acquisition, that is, acquisition from projections, undersampling involves missing some radial directions in the k-space. As will be clarified later, each of these cases requires a specific restoration algorithm, but all the methods face the same problem: to reconstruct an image starting from some collected sparse samples of its Fourier coefficients, while violating the Nyquist rate.

In principle, nothing is known about the sample distribution/shape described for a 2D sample by $f^*(x, y)$. If the image is to be composed of $n \times n$ pixels, the amount of information necessary would be n^2.

For reconstruction from projections, for example, the number of projections required to reconstruct its 2D image (suppose $n \times n$) will be $n_0 \cong \pi n/4$, where n is the number of points sampled in each projection [14], assuming the significant pixels are all included in a circle n pixels in diameter. This implying that all projections have the same length and resolution.

The presence of a number of points in the image with zero value or internal symmetries of the sample can reduce these numbers, but it is necessary to know in advance some sample shape properties. For example, if one knows a target image is circularly symmetric and spatially uniform, as that shown in Figure 5.1, only few views of projections are needed to accurately reconstruct the spin density distribution of the object (for this example, one projection would suffice). Along the same line of reasoning, if a target image is known to be composed by a set of sparsely distributed points, one can imagine that the image can be reconstructed without satisfying the Shannon/Nyquist sampling requirements.

When the positions of significant pixels are not known in advance, an $n \times n$ image can be accurately reconstructed using a data set which has an order of $s \log n$ samples provided that there are only s significant pixels in the image [19] (when this knowledge is available, the number of information required is exactly s).

In the present chapter, these concepts are formalized, prior to a review of sparse sampling and restoration strategies for solving the previous problem. Attention is paid to sparse acquisition methods that not use restoration algorithm (artifacts are tolerated), and a detailed description is dedicated to sparse acquisition methods for which a restoration algorithm is essential.

In the latter class of methods, we differentiate between those that modify the acquisition trajectories (in number or in direction) depending on the sampling pattern (entropy based adaptive methods) and those that adapt to

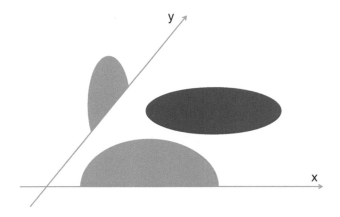

Figure 5.1
Internal sample symmetries, known in advance, allow exact reconstruction from just few projections (in this case, one projection would suffice).

the given acquisition method, trying to reduce artifacts independently of the sampling pattern (compressed sensing).

Moreover, these methods are compared by applying them to some sample images. Comparisons are presented both visually and by calculating the PSNR (defined in Chapter 1).

The chapter concludes with the proposal of some mixed acquisition/reconstruction methods to retain the advantages of the adaptive methods (i.e., to collect "more informative" data) and those of compressed sensing (independence of acquisition strategy and convex optimization reconstruction) and, at the same time, improving each method in image quality and data reduction. Some encouraging preliminary results are reported.

5.2 Sparse Methods without Restoration

Vastly undersampled 3D projections are used to increase temporal resolution and provide better dynamic information for 3D contrast-enhanced angiography (CE)-MRA [6, 87]. Aliasing caused by undersampling in this method often can be tolerated because the vessel–tissue contrast is high. Tsai et al., in [136], propose variable density k-space sampling trajectories to restrict the violation of the Nyquist criterion to the outer part of the k-space.

A method proposed by Lee et al. [68] uses a variable-density stack of spiral trajectories (Figure 3.8(f)) that varies the sampling density both along the $kx - ky$ plane and the kz direction. The method is shown to preserve reasonable image quality while reducing acquisition time by approximately half compared to a fully sampled stack of spirals.

Spiniak et al. [131] use an efficient undersampling method for a progressive guidance trajectory [86]. The main goal of this study is to design an acquisition scheme to cover the k-space as quickly as possible to reduce total scan time during 3D image acquisition. To this aim, a spherical FOV divided into voxels, each of which is centered in a previously defined Cartesian grid position, is defined. The acquisition process consists on the collection of a series of directions starting at the k-space origin and reaching the surface of the FOV after a fixed elapsed time. A list with all voxels is created, sorting them according to their distance to the origin. For every trajectory, the method chooses the nearest untouched voxel from the list, heading toward it. The resulting paths are curve trajectories. The path with the highest efficiency (by maximizing the number of untouched voxels covered between the source and destination in the fixed time) is chosen, compatibly with the constraints regarding the maximum gradient strength and maximum switching rate (or slew rate). After each direction, the touched voxels are removed from the list. As scanning progresses, new trajectories are aimed at regions of the k-space that remain to be sampled. Initially, most of the k-space is unsampled, and hence, the trajectories are very efficient in terms of coverage. For the last trajectories, unsampled regions are very sparse and the process becomes inefficient because the contribution to the total coverage is negligible. The undersampling method is based on reducing the number of trajectories, leaving out those with low new information. This would have limited impact on image quality because the undersampling, giving higher priority to the center region of the k-space, mostly affects the outer regions of the k-space.

The shells method is a non-Cartesian 3D k-space trajectory [54]; fully sampled concentric spherical shells are used for 3D selective RF pulse design [150]. The shells trajectory samples k-space on the surfaces of a series of concentric spheres (Figure 5.2).

A full 3D k-space shells acquisition is implemented in [125], and it is demonstrated to have motion correction properties in [126]. Because the shells trajectory does not acquire data in the corners of k-space, it offers a coverage reduction compared to a fully sampled 3D Cartesian acquisition, thus resulting in a loss of resolution. However, this is substantially larger than the k-space coverage obtained by skipping views in the phase-encoded corners of a Cartesian 3D acquisition, thus resulting in a lower loss of resolution with respect to the Cartesian 3D with skipping views.

As extreme of undersampling, a single k-space shell acquisition has found use as a spherical navigator echo for motion tracking [146].

Shu et al. [127] use to sweep interleaved helical spirals on the surface of

Figure 5.2
Shells k-space trajectories used in [127], coded by continuous and dotted lines, respectively, sweep from one pole to the other of a spherical surface. In an undersampling scheme, some interleaves are skipped, for example, that represented by a dotted line.

each shell from one pole to another (see Figure 5.2). The shells acquisition method is a true center-out trajectory (from the center to the periphery, as the spiral scan) in 3D k-space, in that it starts from the origin of the k-space and extends progressively to the periphery. An important property of the shells trajectory is that a large number of interleaves exclusively collect data from the periphery without crossing the center of k-space. Because the number of interleaves varies on a per-shell basis, there is flexibility in controlling the sampling density as a function of the k-space radius. These properties suggest that the shells trajectory is well suited for undersampling design. The way used to implement undersampling for the shells trajectory method is to remove selected interleaves within a shell. This method is proven to be an optimal balance between acquisition time reduction, undersampling artifacts reduction, and edges preservation.

In PROPELLER (periodically rotated overlapping parallel lines with enhanced reconstruction) [1, 97, 98], data acquisitions consist in a multiple-shot fast spin-echo approach in which several parallel k-space lines are collected in each time of repetition (TR), forming a blade that is then rotated around its center and acquisition is repeated to cover the k-space (see Figure 5.3). The acquisition is radial. PROPELLER is attractive due to its greatly reduced sensitivity to various sources of image artifacts (in particular motion). Arfanakis et al. [1] consider PROPELLER in an undersampled form and study the resulting artifacts.

Other methods utilize information redundancy from temporal correlations of dynamic images as a sort of restoration. Keyhole methods [141, 59] and reduced-encoding MR imaging by generalized series reconstruction (RIGR) [144] assume the temporal variation occurs primarily in the central portion of k-space. For this reason, the keyhole technique is applied in fMRI [134, 35], where resting states (relaxation) alternate to activation states (activity) and, during these states, series of images are collected. To reduce acquisition time, a full k-space reference image is acquired for each state and for subsequent images of the series consists of sampling only a central fraction of the k-space (low k-space data). The missing peripheral lines in the k-space (high k-space data) are supplied from the reference image. In this way, an approximately 60% in acquisition time per image can be saved for T_2^*-weighted gradient-echo schemes, while retaining nearly all the functional information of the original high spatial resolution images [35]. There are, however, two potential problems associated with the keyhole technique that may affect the reconstructed images: noise correlation between different images and discontinuities. Because of data sharing, the high k-space data are the same for all the keyhole images. Therefore, noise between images is correlated. When statistical analysis is performed across a series of images to assess significant activations, the shared k-space data will decrease the effective sampling size and result in a higher noise level.

Regarding discontinuities, the basic principle of the keyhole technique is to combine the dynamic low k-space data with the reference high k-space data. Due to the image-to-image signal fluctuation, amplitude and phase discontinuities may exist at the junction between the reference data and low k-space data. These discontinuities may result in image artifacts and in spatial resolution reduction. A detailed description of the effects of the keyhole method can be found in [152].

Clinical practice shows that a high spatial resolution is essential for most dynamic imaging applications. To this end, some techniques increase temporal and/or spatial resolution by directly exciting the region of interest, such as the zoomed imaging technique [142] thus reducing the data to be acquired.

In seeking the optimal compromise between temporal resolution, spatial resolution, and SNR for a specific application, it is useful to introduce information about the imaging process or the object and so to truly enlarge the amount of information available to aid reconstruction without spending additional measuring time.

(a)

(b)

Figure 5.3
Different PROPELLER *k*-space sampling patterns [1]. In particular, (a) shows
a 6-blades scheme, each blade composed by sixteen lines; (b) shows a 12-blades
scheme, each blade composed by eight lines. Undersampling is realized by
skipping dotted blades.

Reduced field-of-view (rFOV) reconstruction utilizes the fact that only a limited region of an image undergoes substantial variation over time [51, 16]. The idea of rFOV is to exploit the fact that dynamic changes in an image series may be confined to a certain area (the rFOV) within the full FOV. This information has been used to increase temporal resolution in fluoroscopic imaging without sacrificing spatial resolution.

Having been demonstrated first for Cartesian sequences [51], rFOV updating is applicable to any segmented k-space acquisition scheme as radial [119, 145, 95] or spiral imaging [121]. The principal limitation is that erroneous results will be produced if some of the additional information is wrong or the required conditions are not filled, for example, if changes occur to the object outside the rFOV do occur. In clinical practice, it is generally difficult to strictly satisfy such conditions. It is important, therefore, to evaluate the robustness of different implementations of the rFOV technique.

The projection-reconstruction rFOV approach (PR-rFOV) has been demonstrated to be more robust than the other rFOV approaches. The greater robustness is a consequence of the special aliasing properties of projection-reconstruction imaging and is considered to be crucial for clinical applications (as will be clarified below).

Both keyhole and rFOV methods are based on the fact that the information from relatively stationary regions in object space is considered to be redundant.

The UNFOLD method (unaliasing by Fourier-encoding the overlaps using the temporaL dimension) [76, 137] supposes data have to be collected both in spatial and in temporal directions; the same image has to be reconstructed more times to reproduce temporal variations occurring inside it.

UNFOLD uses the t axis of $k-t$ space (the combination of k-space with a time axis [151]) to resolve information normally encoded along the k axis. In applications in which the time axis is not efficiently exploited by conventional encoding, such a reorganization of $k-t$ space can lead to a significant decrease in acquisition time for the temporal frames. In this case, spatially aliased versions of the same image (each image having a determined phase shift offset) are collected; spatial aliasing is then eliminated by separating overlapped information through a temporal Fourier transform before the spatial fast Fourier transform. This can lead to a temporal resolution improvement by nearly a factor of two in cardiac-triggered imaging, and by as much as a factor of eight in fMRI.

Depending on the situation, this acquisition time reduction per temporal frame can be translated into a reduction of the total imaging time, an improvement of the spatial or temporal resolution, or an increase in the spatial coverage. The reduction can also be used to allow a faster pulse sequence to be replaced by a slower one while preserving time resolution. This method cannot be used to improve spatial resolution when temporal data are not necessary or available.

Temporal acceleration can be achieved with the $k-t$ Broad-use Linear Acquisition Speed-up Technique ($k-t$ BLAST), which uses training data as prior information along with the interleaved sampling function [138]. Conceptually, UNFOLD and $k - t$ BLAST reduce information redundancy by economically using $k - t$ space.

Originally, $k-t$ BLAST has been defined for Cartesian k-space trajectories [138, 151], due to the simplicity of describing and correcting for the aliasing produced by rectilinear sampling patterns. It has since been presented as a non-Cartesian sampling/reconstruction method [44]. Spiral readouts, in particular, are appropriate for some applications [84] because they have good flow properties and are time efficient, allowing further increases in spatial and/or temporal resolution when combined with undersampling techniques. Shin et al. [124] elegantly solve the problem for spiral trajectories. An improvement to the UNFOLD method is proposed in [53] where UNFOLD is combined with the undersampled 3D stack of spirals acquisition [52] to increase spatial coverage for high resolution fMRI.

5.3 Sparse Methods with Restoration

5.3.1 Sample Adaptive Acquisition/Restoration Methods

When imaging from projections is used, some undersampling approaches exist, based on adapting the directions to be collected to the sample shape. As described above, in the absence of preliminary information about the sample shape, $f^*(x,y)$, the projections are usually sampled at a regular set of angular positions. For a 2D reconstruction, these will be measured at constant increments of at least π/n_0 radians (to cope with noisy data, this minimal number is usually exceeded). This is an obvious choice because nothing is known about the sample shape and symmetries until image reconstruction has actually been carried out. As first example of adopting the acquisition process based on the sample shape, Contreras et al. [23] proposed a sparse sampling method for MRI nondestructive analysis of wooden logs, taking advantage of cylindrical symmetry by acquiring transverse 1-D projections with a helical undersampled pattern. Linear interpolation is used to estimate the skipped data, and slice images are reconstructed by filtered back-projection. The sequence is improved by using selective multipass scanning, without major variations of the scan time. While the technique is particularly useful for logs, since they present several characteristics that can be used to reduce the long scan time, one can imagine applying similar strategies for diagnostic purposes. In what follows, we discuss an entropy-based adaptive acquisition method from projections in details.

5.3.1.1   Entropy-Based Adaptive Acquisition Method

Placidi et al. [103, 105] present an adaptive acquisition technique for MRI from projections, first defined in the image space [103] and then in k-space [105], to reduce the total acquisition time by collecting only the most informative projections, without any a priori information about the sample. The adaptive method exploits information present in the projections that have been collected during the acquisition process to identify the sample symmetries if any.

    This is possible through the calculation of a function, called the *entropy*, which is a measure of information content of each projection and allows differences in information content between consecutive collected projections to be identified during the acquisition progress. This is then made to "sparsify" the projections where high similarity between projections is present. In the image domain, the entropy can be used to elucidate sample internal symmetries, smoothness, or regular shape. In the k-space method [105], the entropy function is defined on the power spectrum of the projections. In what follows, we describe this method (some material is reproduce with kind permission from Elsevier[1]).

    The process starts by measuring the projections at four fixed orientations: 0, 45, 90, and 135. The evaluation of the difference in information content of these four initial projections is performed, followed by the selection of new angles where the difference in information content is maximum. The next projection is measured between the two where the information difference has a maximum. The entropy of the new projection is calculated, and the selection of the subsequent projection is made. The procedure is repeated until the maximum difference in information content between projections falls below a data-driven (calculated) threshold ϵ.

    Algorithm 5.1 describes the acquisition method in a general form.

    The method allows the total acquisition time to be reduced, with little degradation of the reconstructed image, adapting itself to the arbitrary shape of the sample. The choice of, approximately, the most informative projections is made, taking into account the information content of the previous projections. To apply correctly this method, two functions, the entropy (E) and the difference of information content (ΔInf), and one parameter, ϵ, are necessary; we briefly introduce them.

    In Chapter 2, we introduced the concept of projection $p_\phi(r)$ of a function. In this case, we can remark that an $n \times n$ function $f^*(x, y)$ is reconstructed from experimental data measured at a limited number of projections, each sampled at a fixed number of points, n. Without loss of generality, we can consider the integral of $f^*(x, y)$ (that is, of each of its projections) can be

---

[1] ω-space adaptive acquisition technique for magnetic resonance imaging from projections, G. Placidi, M. Alecci, and A. Sotgiu, *Journal of Magnetic Resonance*, 143(1), 2000, 197–207.

Algorithm 5.1: Calculate the, approximately, most informative set of projections: in short

---

Collect the initial set of four projections
For each projection ϕ, calculate the difference in information content with the previous projection, ΔInf_ϕ, and store it in the vector ΔInf
Calculate ϵ
while $max(\Delta Inf) \geq \epsilon$ **do** {The acquisition terminates when the maximum difference in information is below ϵ}
    Extract the projections ϕ_- and ϕ_+ yielding $max(\Delta Inf)$
    Collect a new projection, at ϕ_p, between ϕ_- and ϕ_+
    Update the vector ΔInf to contain the differences between ϕ_p and ϕ_- and between ϕ_p and ϕ_+
end while
Return the measured set of projections

---

normalized to $1/n$. The dc value of the FT of each projection is then exactly equal to $1/n$, and the amplitude of any other frequency component of each projection is less than, or equal to, $1/n$. This condition will be used in defining the information content of each projection. Let $P_\phi(k)$ be the power spectrum of $p_\phi(r)$ at the angle ϕ.

When data are obtained by sampling a physical process, noise is always present. We suppose that the real and imaginary components of the FT of the projection are corrupted by the presence of white noise of variance σ^2 and zero mean value. In this case, the power spectrum $P_\phi(k)$ is defined as

$$P_\phi(k) = R_\phi^2(k) + n_{R_\phi}^2(k) + 2R_\phi(k) \cdot n_{R_\phi}(k) + I_\phi^2(k) + n_{I_\phi}^2(k) +$$
$$+ 2I_\phi(k) \cdot n_{I_\phi}(k) = P_\phi(k) + Pn_\phi(k) + g(P_\phi(k), Pn_\phi(k)), ... \tag{5.1}$$

where $n_{R_\phi}(k)$ and $n_{I_\phi}(k)$ are the noise terms contained in the real and imaginary parts (R_ϕ and I_ϕ) of the measured projection (at the angle ϕ), respectively, $Pn_\phi(k)$ is the power spectrum of noise in the $k-th$ point (taking into account the noise affecting both real and imaginary parts) and $g(P_\phi(k), Pn_\phi(k))$ is a function including the products between the noise terms and the signal terms.

To evaluate the information content of different projections, we introduce the "entropy" $E(P_\phi)$ of a projection defined in terms of its power spectrum:

$$E(P_\phi) = \sum_{k=1}^{n} P_\phi(k) log \left(\frac{1}{P_\phi(k)} \right). \tag{5.2}$$

The contribution $P_\phi(k) log \left(\frac{1}{P_\phi(k)} \right) = 0$ for $P_\phi(k) = 0$ and for $P_\phi(k) = 1$.

As base of the logarithm, we take the square of the number of points sampled for each projection (n^2), such as $E(P_\phi)$ assumes values between $1/n^2$ and $1/n$.

$E(P_\phi) = 1/n^2$ corresponds to the case in which the power spectrum of the projection is zero everywhere with the exception of the dc component at which its value is $1/n^2$. The projection has constant value everywhere, and it has the minimum information content.

When all the points of the power spectrum of the projection have the same value, equal to $1/n^2$, then $E(P_\phi) = 1/n$ and the projection has the maximum information content. The projection in this case corresponds to a Dirac δ function.

A projection is considered to be more significant when it gives more details of the distribution of $f^*(x, y)$ and, as seen before, this occurs for greater values of the entropy $E(P_\phi)$. The function $E(P_\phi)$ gives an estimate of the information content that is independent of the observer; it only depends on the shape of the object to be observed.

To include the effect of noise in the definition of entropy, we can assume that $P_\phi(k)log(1/P_\phi(k)) = 0$ for $P_\phi(k) \leq 2(m_{pw} + \sigma_{pw})$ and for $P_\phi(k) \geq 1/n^2 - 2(m_{pw} + \sigma_{pw})$, where m_{pw} and σ_{pw} are the mean value and the standard deviation of the noise (defined by the square of the Gaussian noise present in the measured terms, real and imaginary), respectively. The factor 2 is due to the sum of the two noise terms that are assumed to have the same power value and characteristics. It can be shown [39] that m_{pw} and σ_{pw} are related to the standard deviation of the original noise by the following relationships:

$$m_{pw} = \sqrt{\pi/2}\sigma, \qquad (5.3)$$

$$\sigma_{pw} = \sigma\sqrt{2 - \pi/2}. \qquad (5.4)$$

In this way, the terms exclusively due to the noise can be compensated but not those due both to signal and noise as $g(P_\phi(k), Pn_\phi(k))$.

The function used to select the new projection from the ith and $(i + 1)$th is

$$\Delta Inf_i = \Delta E_i \cdot \Delta \phi_i \cdot K_i, \qquad (5.5)$$

where $\Delta E_i = |E(P_{\phi_{i+1}}) - E(P_{\phi_i})|$, $K_i = max(E(P_{\phi_{i+1}}), E(P_{\phi_i}))$, and $\Delta\phi_i = |\phi_{i+1} - \phi_i|$. The factor K_i guarantees that in cases where multiple projection pairs share parity of difference in entropy, the algorithm selects projections having greater entropy, namely, higher information. The term $\Delta\phi_i$ guarantees that the angular difference between acquisitions is not too high, reducing the chance that important projections are missed. This could otherwise occur in the proximity of the minima and maxima of the entropy.

The minimum difference ϵ at which the process of selection of the projections ends is determined experimentally by choosing ΔInf_i as that evaluated between the projection that has the maximum entropy, of the first four initially collected, and a fifth projection measured at $4/n$ rad from it. This is done just before the acquisition starts. In the presence of noise, we have to consider that the function ΔInf_i is proportional to the sum of three terms, one accounting for the difference in information between two projections (A), another being principally due to the noise (B), and one which is a mixture of signal and noise (C). In fact, when noise is considered, the expression for ΔE_i becomes

$$\Delta E_i = |\sum_{k=1}^{n}\{[P_{\phi_{i+1}}(k) + Pn_{\phi_{i+1}}(k) + g(P_{\phi_{i+1}}(k), Pn_{\phi_{i+1}}(k))]\cdot$$
$$\cdot log(T_{i+1}(k)) - [P_{\phi_i}(k) + Pn_{\phi_i}(k) + g(P_{\phi_i}(k), Pn_{\phi_i}(k))]\cdot$$
$$\cdot log(T_i(k))\}| = |A + B + C|, \tag{5.6}$$

where $P_{\phi}(k)$, $Pn_{\phi}(k)$, and $g(P_{\phi}(k), Pn_{\phi}(k))$ are defined as in Equation 5.1, $T_{i+1}(k) = [1/P_{\phi_{i+1}}(k) + Pn_{\phi_{i+1}}(k) + g(P_{\phi_{i+1}}(k), Pn_{\phi_{i+1}}(k))]$ and $T_i(k) = [1/P_{\phi_i}(k) + Pn_{\phi_i}(k) + g(P_{\phi_i}(k), Pn_{\phi_i}(k))]$, the indices $i+1$ and i referring to different projections, and the terms A, B, and C are defined as

$$A = \sum_{k=1}^{n}\{P_{\phi_{i+1}}(k)\cdot log(T_{i+1}(k)) - P_{\phi_i}(k)\cdot log(T_i(k))\}, \tag{5.7}$$

$$B = \sum_{k=1}^{n}\{P_{\phi_{i+1}}(k)\cdot log(T_{i+1}(k)) - P_{\phi_i}(k)\cdot log(T_i(k))\}, \tag{5.8}$$

$$C = \sum_{k=1}^{n}\{g(P_{\phi_{i+1}}(k), Pn_{\phi_{i+1}}(k))\cdot log(T_{i+1}(k)) +$$
$$- g(P_{\phi_i}(k), Pn_{\phi_i}(k))\cdot log(T_i(k))\}. \tag{5.9}$$

Lowering of the SNR causes an increase in B and C and, as a consequence, an increase in the oscillations in ΔInf_i, that is of the uncertainty of the calculated value of ΔInf_i.

To set a value for ϵ, we make the following assumptions.

1. The term K_i of Equation 5.5 is taken to be equal to $1/n^2$, which is the minimum value it can assume.

2. The minimum angular distance is taken to be equal to $4/n$ rad. ΔE_i is taken to be equal to the difference in entropy between the two considered projections[2].

In this way, we take into account both the minimum angular distance between projections and the minimum difference of entropy between projections.

To avoid having the noise in the projections affect the value of ϵ, we subtract from the estimated value of ΔE_i an estimate of B. This is done by substituting in Equation 5.8 the level of noise given by $2(m_{pv} + \sigma_{pv})$.

---

[2]As explained above, these form the pair between the projection that has the maximum entropy value, of the first four initially collected, and a fifth projection measure at the minimum angular distance, $4/n$ rad, from it.

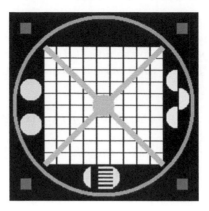

Figure 5.4
Digital 256×256 image of a numerical phantom.

The term C (Equation 5.1) cannot be estimated, being a function both of signal and of noise. The calculation of ϵ ensures that the minimum number of projections does not change for different SNR values.

In the case of images of complex structure, the number of collected projections approximates the estimated minimum number of projections, m_0. The number of projections becomes lower when the image has internal symmetries or, being smooth, it is mapped at fewer levels. Using a conservative value for ϵ guarantees that the image is reconstructed without any loss of information.

A detailed description of the adaptive method is sketched by Algorithm 5.2.

5.3.1.2 Adaptive Acquisition Results

The above adaptive k-space acquisition method has been tested on simulated data using the MRI simulator discussed in Chapter 2. The test image (Figure 5.4) considered was 12.8×12.8 cm^2, having a resolution of 0.5 mm. We fixed the main field to 0.1 T, $T_1 = 0.8$ s, and $T_2 = 0.16$ s for each pixel. To preserve the original dimensions in the reconstructed image, we choose the following acquisition parameters: sampling period 40 ms, sampling rate 6.4 KHz, gradient value 0.12 G/cm, and a rectangular pulse shape of 10 ms in duration. A set of 500 k-space projections (each sampled on 256 points) of the phantom was generated.

The actions of the pulse sequence and of the gradients were defined in a sequence table. The program calculated the real and imaginary parts of the 500 projections, in steps of $0.36°$, in the xy plane of the image, described by using the appropriate xy gradient combination.

Without loss of generality, projections were collected assuming the mag-

---

Algorithm 5.2: Calculate the, approximately, most informative set of projections: in details

---

for $i = 1$ to 4 **do** {For the starting set of projections}
    Collect projection ϕ_i
    Calculate Entropy $E(P_{\phi_i})$
    $EP[i] \leftarrow E(P_{\phi_i})$
end for
for $i = 1$ to 4 **do** {For the starting set of projections}
    Calculate differences of information content ΔInf_i
    $\Delta Inf[i] \leftarrow \Delta Inf_i$
end for
{Find the maximum value of information difference and its index}
$MAX \leftarrow max(\Delta Inf)$
$s_{max} \leftarrow index(max(\Delta Inf))$
{Measure a new projection where the entropy is maximum and calculate ϵ}
$s \leftarrow index(max(EP))$
Collect a 5th projection $4/n$ rad close to the sth
Calculate ϵ
while $MAX > \epsilon$ **do** {The acquisition terminates when the maximum difference in information is below ϵ}
    Collect a projection at $\phi_p = \frac{\phi_{s_{max}} + \phi_{s_{max}+1}}{2}$
    Calculate $E(P_{\phi_p})$
    Calculate ΔInf_p
    Calculate ΔInf_{p+1}
    {Update the vectors EP and ΔInf to the position corresponding to the angle ϕ_p}
    Insert $E(P_{\phi_p})$ into the vector EP
    Insert ΔInf_p into the vector ΔInf
    Insert ΔInf_{p+1} into the vector ΔInf
    {Update the maximum value of information difference and its index}
    $MAX \leftarrow max(\Delta Inf)$
    $s_{max} \leftarrow index(max(\Delta Inf))$
end while
Return the measured set of projections

---

netization was completely recovered between acquisition of consecutive projections in the absence of noise.

Different subsets of projections were extracted using either the adaptive criterion or the regular acquisition method. Fourier reconstruction (with nearest neighbor interpolation to fill missing Fourier coefficients, corresponding to missing projections) was used to obtain the reconstructed images with a low number of projections. The images were compared both visually and numerically by calculating PSNR (though noiseless data were used, noise was produced after reconstruction, in the form of artifacts, due both to the reconstruction process and to the limited number of projections used).

In this example, the image used for comparison is that reconstructed by using the whole set of projections. The comparison with the image obtained with the whole set of projections, and not with the theoretical starting image, is necessary to eliminate the effects of the reconstruction method from the calculated error of the image under inspection.

The reference image, obtained by using the whole set of projections, and the images obtained by using different sets of projections are reported in Figure 5.5.

In particular, reconstructions obtained by using the whole set of 500 projections, a set of 120 FIDs (whose orientations are calculated with the adaptive method), a set of 125 FIDs taken at regular angular steps, and a set of 201 projections taken at regular angular steps are shown in Figure 5.5a–d, respectively. The number of the projections, 120, was calculated by the adaptive acquisition method and depends on the shape of the original test image (whose information is collected on the difference between the information content of the projections).

The number of projections arrived at with the adaptive algorithm is about 60% of the minimum number of projections, 201, necessary to reconstruct an image having the dimensions and resolution of the given test image, without any a priori information of the sample itself.

Figure 5.5(b) resembles the target image very well, although it is obtained with 12% of the total number of projections and with five projections less than the set used to reconstruct Figure 5.5(c). In particular, Figure 5.5(b) does not show noticeable artifacts.

Figure 5.5(c) clearly shows some artifacts on both the external and the internal features of the image, due to the low number of projections used in the reconstruction.

The differences between the two undersampled images are also evident from the PSNR values (see the caption of Figure 5.5); the adaptive reconstruction is very similar to that of the completely regularly sampled image though it has been obtained with 60% of the required projections. The undersampled regularly spaced projections on the other hand yielded a PSNR 1/3 lower. In order to better show the artifacts due to angular undersampling,

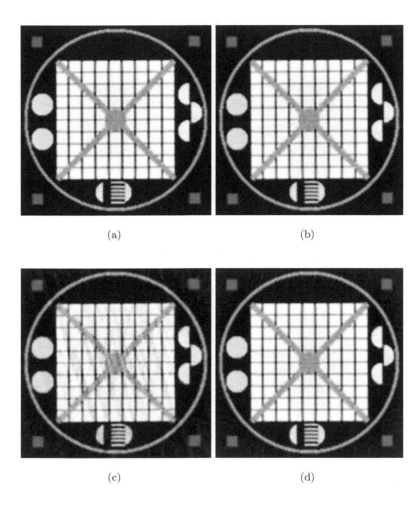

(a) (b)

(c) (d)

Figure 5.5
Reference image (a) reconstructed using the whole set of 500 simulated projections. The images reconstructed using (b) 120 projections collected by the adaptive method and by regularly spaced projection reconstruction (c) undersampled at 125 and (d) fully sampled with 201 projections. PSNR values, with respect to (a), are: 37.21 dB (b); 23.86 dB (c); 36.57 dB (d).

we report, in Figure 5.6, the differences relative to the target in an expanded gray scale.

The adaptive method allows the acquisition of a near-optimal set of projections close to the most informative set of projections composed by the minimum number of projections, but not the optimal one. In fact, though very effective in reducing the acquisition time and undersampling artifacts, this method suffers from the following limitations; some important projections can be excluded from the acquired set, especially in the proximity of ΔInf minima or maxima. Some redundant projections can be collected, especially in the proximity of ΔInf sharp variations.

For the example image in Figure 5.4, the plot of ΔInf is reported in Figure 5.7. The plot highlights symmetries of the sample and the rapid variation of ΔInf; it can be very difficult to measure exactly over the maxima (minima) of ΔInf. For this reason, the presented method gives an approximation of the optimal set of projections.

5.3.1.3   An Adaptive Acquisition Improvement

In [100] (some material is herein reproduced with kind permission from Springer+Business Media[3]), Placidi considers the problem of measuring exactly the most informative set of projections by collecting a priori information about the sample through preliminary measurement of circular paths at different distances from the k-space center (such circular paths have been used elsewhere as the basis of motion correction [146]). The idea behind the algorithm is to consider how the power spectrum of a standard MR image (Figure 5.9) is mainly distributed along radial k-space paths. In some directions, the distribution terminates before the k-space border has been reached. In others, the distribution does not start from the k-space center and extend to the k-space border. In order to take into account these opposing situations, a set of preliminary circular trajectories are collected. The circular trajectories allow the interception of the most important projections. By analyzing the data from these trajectories, it is possible to establish the desired set of projections before the image acquisition starts.

The acquisition process consists of the preliminarily collection of S concentric circular trajectories having the center in the image k-space center. The directions of the most informative projections are then set by using the information about the power spectra of these paths. In particular, from the logarithm (used to reduce differences in scale) of the power spectra of the collected trajectories, the mean values are calculated, and maxima above the mean value, in both curves, are used to indicate the presence of the most

---

[3]Circular acquisition to define the minimal set of projections for optimal MRI reconstruction, G. Placidi, *Lecture Notes in Computer Science, Computational Modelling of Objects Represented in Images*, 6026, 2010, 254–262.

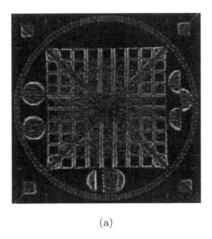

(a)

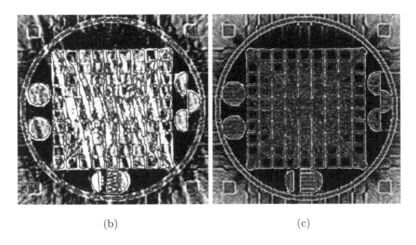

(b) (c)

Figure 5.6
Difference images obtained by subtracting Figure 5.5(a) from Figure 5.5(b) (a); Figure 5.5(c) (b); Figure 5.5(d) (c). Images are shown in an expanded gray scale.

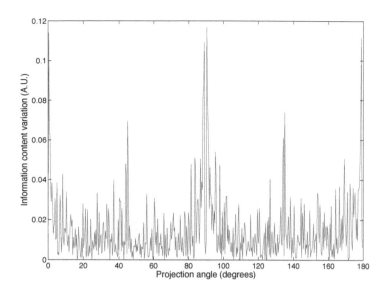

Figure 5.7
ΔInf plot for the projections of the image shown in Figure 5.4. Internal sample symmetries produce a symmetrical plot of ΔInf.

informative radial directions. The flowchart of the method is summarized in Figure 5.8. To describe the process, we first note that L and H are two matrices. In particular, L contains two columns: the first is a list of the angles at which measurements are performed; the second is a binary indicator of whether that angle was chosen by the algorithm. Initially set to 0, at the end of the analysis, the chosen angles will have 1 in that column. H represents the matrix where the S circular paths are stored in columns.

The samples along each circular path are measured at the same angles, so the number of rows of L and H is N, one angular value for each sample. N is calculated to consider the width of a pixel situated on the circle of maximum radius as the minimum angular separation, that is, $\Delta\alpha = 1/(Rs_{max})$, where Rs_{max} is the radius of the maximum circle. In this way, Nyquist criterion is maintained for the external circular path.

The vector M contains the mean values of the logarithm of the power spectra of the measured circular paths, and it has length S. The output of the algorithm, namely, the set of the chosen angles, is used by a standard acquisition sequence from projections to collect the necessary data set. It is important to note that:

1. data measured to collect information about the sample (related to

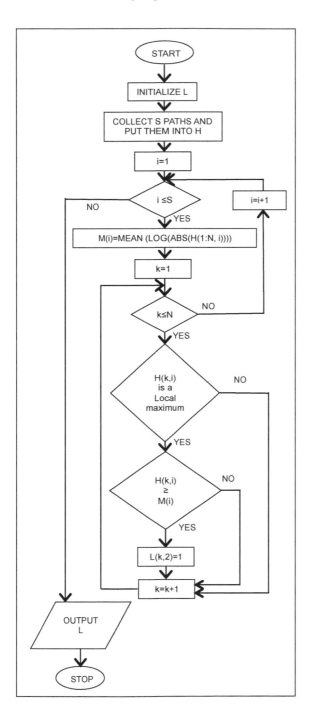

Figure 5.8
Flowchart diagram of the modified adaptive method.

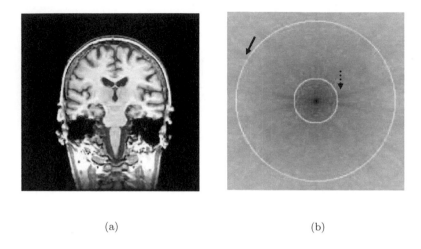

(a)   (b)

Figure 5.9

A completely sampled (256×256) MRI image used as a test (a) and its k-space power spectrum (b), represented in logarithmic form; the two circles indicate the circular trajectories used to collect preliminary imaging information. Arrows indicate the presence of high power zones. To avoid saturation, (b) is shown in inverted mode (darker values correspond to greater values).

      the circular trajectories) can also be used in reconstructing the final image, thus improving its quality;

2. the algorithm requires some preliminary time to collect the necessary information about the optimal angle set, before the acquisition of radial projections starts.

The termination parameter used to limit the number of projections for the $i - th$ circle is represented by $M(i)$. All local maxima above this value are chosen; the others are discarded. Each circle has its own termination parameter (this happens because the power spectrum decreases with distance from the center). The set of these angular directions is considered to be the optimal, most informative, set of projections to be collected.

    As an example, consider the image shown in Figure 5.9(a). It is a complete MRI, 256×256, coronal image of the head of a healthy volunteer collected with a commercial 1.5 T scanner, from which k-space data have been obtained through inverse Fourier transform. A completely sampled image was chosen to allow numerical calculation of a complete set of radial projections.

    The optimal subset of projections was extracted using the proposed method with $S = 2$. Samples along two circular trajectories at 0.2R and

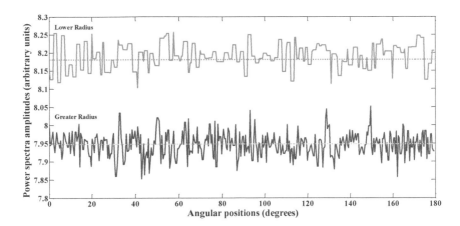

Figure 5.10
Power spectrum plot of the two circular paths outlined in Figure 5.9(b). The corresponding mean values are represented by dotted lines.

0.9R were first acquired. The circles indicated in white in Figure 5.9(b) show the trajectories, while Figure 5.10 shows the power spectra of the two paths, with the corresponding mean values.

Subsets of the projections were extracted using the proposed method and the previously described blind adaptive criterion. The modified Fourier reconstruction algorithm, including an interpolation method [102], was used in both cases to obtain the reconstructed images with a low number of projections. The images have been compared both visually and numerically by using the PSNR calculated between the image to be tested and the image reconstructed by using the whole set of projections.

Also in this case, the comparison with the image obtained with the whole set of projections, and not with the theoretical starting image, was made to eliminate the effects of the reconstruction method from the calculated error of the image under inspection.

The target image, obtained by using the whole set of projections, is shown in Figure 5.11. The images obtained by using a set of 89 projections, obtained by the application of the proposed method, and by using a set of 106 pro-

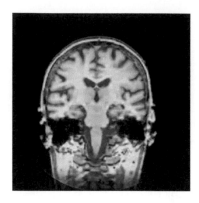

Figure 5.11
Image reconstructed by collecting a complete set of 500 radial projections from
the test image of Figure 5.9.

jections whose orientations were calculated by applying the blind adaptive
acquisition method, are shown in Figures 5.12(a) and 5.12(b), respectively.

The number of projections obtained with the blind method, 106, depends
on the shape of the original test image. It is important to note that the
numbers obtained by the two methods are not so different, but the blind
adaptive acquisition method collects a greater number of projections because
it acts without any a priori information about the sample. Moreover, the
sets of projections obtained by the two adaptive methods shared a subset of
84 projections: 5 projections were specific to the proposed method, and 22
projections were specific to the blind method (many of them were those used
to collect initial information for the blind adaptive method).

The estimated number of projections is 40% and 53% of the minimum
number of projections, 201, necessary to reconstruct an image having dimensions and resolution of the given test image.

The images obtained by the two methods are quite similar. In fact, they do
not show noticeable artifacts because of the optimal and near-optimal choice
of the angular positions of the collected projections. The similarities between
the two images are also demonstrated by the values of PSNR.

Difference images between Figure 5.11 (image reconstructed by a complete
set of projections) and Figures 5.12(a) and 5.12(b), in modulus, are shown
in Figure 5.12(c) and Figure 5.12(d), respectively, in an expanded gray scale,
demonstrating the similarities and the lack of undersampling artifacts (in
the bottom region, Figure 5.12(d) shows a low residual of the original image
profile).

The present method produced a further 20% reduction in the number of

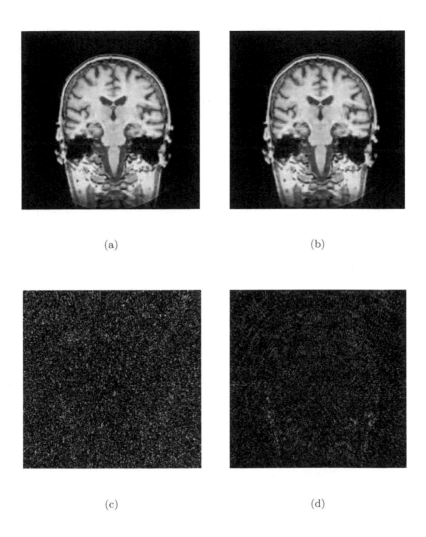

(a)　　　　　　　　　　　　(b)

(c)　　　　　　　　　　　　(d)

Figure 5.12
Images reconstructed by collecting a set of 89 projections determined by the modified adaptive method (a) and a set of 106 projections collected by using the blind adaptive method (b). PSNR = 33.1 dB for Figure 5.12(a) and PSNR = 31.92 dB for Figure 5.12(b). (c) and (d) report the images obtained as the modulus of the differences between Figures 5.11 and 5.12(a) and 5.12(b), respectively; the gray scale of the difference images has been expanded to highlight lower details.

collected projections with respect to the blind method. Obviously, this method required the collection of the two circular paths before data acquisition, and it has an arbitrary base. In fact, the number and choice of the radii of the additional circular paths was based on experimental trial and error deduction. In spite of this, the method allows to save acquisition time with respect to blind method; it is completely deterministic and follows the shape of the imaged object whatever it is.

Conversely, the blind method has two main drawbacks: some information maxima cannot be collected; optimal calculation hardware/software are required for a real-time implementation. The method maximizes the image information content while reducing the number of collected projections.

In fact, when described in Placidi et al. [105], the blind adaptive method was compared with the regular acquisition and the result was in favor of the adaptive method.

Though they require some preliminary time to collect the necessary information about the optimal angle set before the acquisition of radial projections starts, the adaptive algorithms allow both an improvement in image quality and a reduction of the number of k-space coefficients with respect to other, nonadaptive, methods.

The adaptive methods proposed in [103, 105, 100] use a restoration/reconstruction method based on FT and nearest-neighbor interpolation [104]. Nearest-neighbor interpolation is justified by the fact that close measured projections are very similar because of the adaptive acquisition methods used.

5.3.2 Sample Independent Acquisition/Restoration Methods

We now consider some undersampling methods applicable where the redundancy is inherent in the image but the image properties (for example, sparseness or symmetries) are not used while performing data acquisition.

Placidi et al. [110] describe an algorithm which is effective in reducing truncation artifacts due to missing k-space samples in MRI. The algorithm works first by filling the incomplete matrix of coefficients with zeroes and then adjusting, through an iterative process, the missing coefficients by reducing the undersampling artifacts.

This set of coefficients is then used as a basis for a superresolution algorithm that estimates the missing coefficients by modeling the data as a linear combination of increasing and decreasing exponential functions, through Prony's method [7]. Prony's method consists of the interpolation of a given data set with a sum of exponential functions; the MRI signals are well represented as a sum of exponential functions, and the missing data can be extrapolated by this representation.

The algorithm performs very well, both for missing rows in Cartesian sampling schemes (phase-frequency acquisitions) and for missing angles in radial schemes (acquisition from projections) undersampling, but it requires some computational overhead. A simpler variation of this method is reported in [101] where a constraint for iterative reconstruction, capable of dealing with any sparse acquisition method, is introduced.

The methodology suggested below is based on the attempt to fill in the missing complex k-space values iteratively, making the assumption that the image has to be zero outside a compact support. This approach transforms the original problem into an interpolation problem in the complex domain. The novelty is that it deals with iterative interpolation in k-space based on the elimination of the artifacts from an extended support of the reconstructed image.

The results, simulating different sparse acquisition strategies (Cartesian, radial, and spiral sampling), are not significantly different from those obtained from other, more complicated, iterative methods (e.g., [110]). Nevertheless, to interpolate in the frequency domain can be very delicate, due to the fact that closeness of Fourier terms does not implies their values are similar, as will be discussed below.

5.3.2.1 Compressed Sensing

During the last few years, the emerging theory of compressive (or compressed) sensing (CS) [19, 29, 20] has offered great insight into both when and how a signal may be recovered to high accuracy (or, for some instances, exactly) even when sampled significantly below the Nyquist rate. In order to obtain this, CS theory needs the signal to be:

1. Sampled on randomly chosen positions

2. Sparse in some basis (transform sparsity)

3. Reconstructed by using nonlinear methods

Random sampling ensures incoherence of undersampling artifacts; aliasing produced by sampling below the Nyquist's limit then has a distribution that appears as uncorrelated noise rather than defined, specific, deterministic pattern that characterizes regular sampling.

To highlight the importance of random sampling in reducing aliasing, we use a simplified version of the 1D signal reported in [74], illustrated in Figure 5.13. A sparse signal is undersampled in the k-space domain (after the FT of the original signal) using both random (indicated by the upper row of dots in Figure 5.13b) and regularly spaced pattern of samples (lower row of dots in Figure 5.13b). Then, the missing values are zero-filled, and the FT^{-1} is calculated to obtain the respective signals. As can be noted, in the random sampling case, aliasing is distributed like random noise (Figure 5.13c), while for the regular undersampling (Figure 5.13d), aliasing completely replicated the signal. In this latter case, determining the true location is not possible. The random noise appearance on the other hand would allow a simple threshold to be used to localize the signals (Figure 5.13e,f). This represents a heuristic way to obtain a perfect reconstruction of the original signal from random undersampling.

The sparsity of an image refers to the number of its significant "coefficients" in some representation (these coefficients are the pixels if the image is analyzed in its original domain, are the Fourier coefficients if it is analyzed in the frequency space, etc.); an image is considered to be s-sparse in some domain if only s of its coefficients have values above a fixed threshold (in most cases, this threshold is zero). The transform sparsity condition indicates that if the image to be reconstructed has a sparse representation in some domain (that is the image is compressible in some domain), it can be perfectly reconstructed by sampling at a significantly smaller rate than the Nyquist rate. In the previous example, the proposed signal clearly had a sparse representation.

Nonlinear reconstruction is a good method to force both transform sparsity and consistency of the reconstruction with the collected data. In application to MRI, it is necessary to explore these aspects more deeply. The fruitful application of CS in MRI critically relies on the answers to the following questions:

1. Can MRI data be collected by using random sampling?

2. Are MRI images sparse?

3. If an image is not sparse, can we use some transforms to make it sparse?

Regarding the first question, MRI acquisition can indeed be designed to achieve pseudo random undersampling. Figure 5.14 shows some usable MRI acquisition trajectories that incorporate a random component and the relative Point Spread Functions (PSF) [74]. The PSF for a given sampling scheme is obtained as Fourier transform of the binary Cartesian matrix having 1 where k-space was sampled and zero elsewhere (zero filling). For complete Cartesian sampling, the PSF corresponds to a δ function positioned in the central point of the image. When an undersampling scheme is used, the PSF leaks energy from the center to the outer points as blurring or artifacts. The maximum leakage level, measured as the maximum value of the PSF excluding the center, gives a measurement of the amount of coherence the sampling method can achieve [74]. For a regular undersampling scheme, the leakage is more concentrated and, as a consequence, the coherence is high. On the contrary, for an irregular undersampling scheme, the leakage is distributed almost uniformly and the coherence is low. A completely random sampling has the lowest coherence level, that is, the necessary condition for CS application has completely filled. This is the case of 3D acquisition (Figure 5.14b) where minimum coherence is obtained in those planes coded by two phase-encoding gradients (the third dimension is coded by using a readout gradient). However, this technique

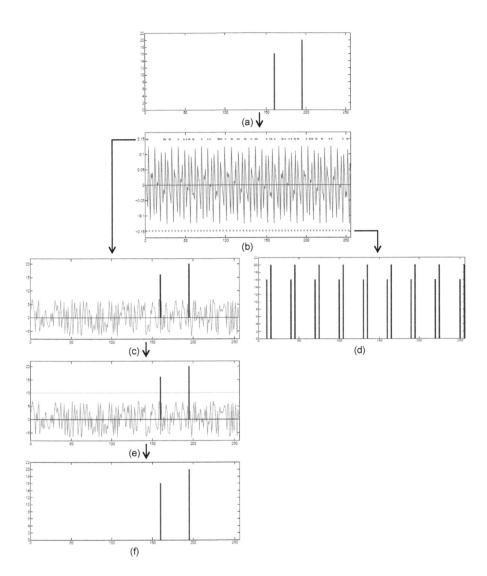

Figure 5.13
A signal (a) is sparsely undersampled (b) both in random way (upper part
dots) and in regular way (lower part dots). Random undersampling aliasing
(c) is in the form of random noise and can be eliminated through thresholding
(e) to obtain a correct signal recovery (f). Regular undersampling aliasing (d)
is in the form of unavoidable repetitions. Similar to Figure 5 in [74], different
data were used.

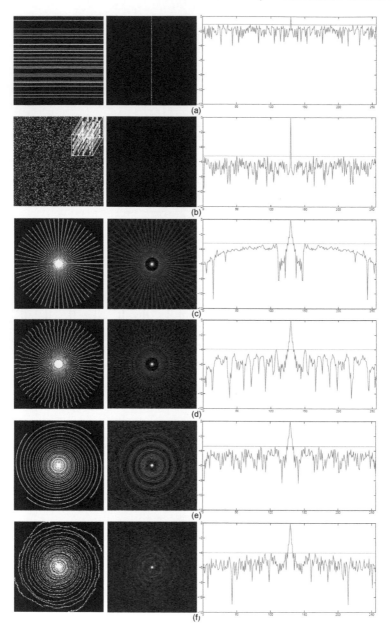

Figure 5.14
Sampling trajectories (left): (a) random parallel lines, (b) random points in a cross section of random lines in 3-D, (c) uniform radial lines, (d) perturbed uniform radial lines, (e) variable density spirals, and (f) variable density perturbed spirals. PSFs (center) and PSFs most significant directions (right). Horizontal line indicates the coherence level. Similar to Figure 6 in [74], different data were used.

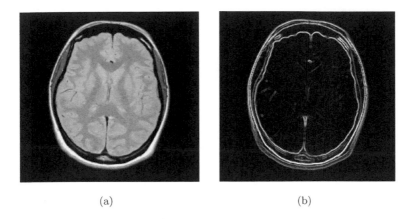

(a)   (b)

Figure 5.15
A sample MRI brain image (a) and its gradient image shown in magnitude form (b).

necessitates collecting a 3D data set so imposing a low acquisition time if volumetric data is not required. Other 2D acquisition strategies (those reported in Figures 5.14a, 5.14c, and 5.14e) following approximately regular paths in the 2D k-space (random walks are impossible) have higher coherence; that is, a reduced level of randomness can be ensured. Some gain can be obtained by perturbing variable density radial projections (Figure 5.14d) or by perturbing variable density spirals (Figure 5.14f). This will ensure pseudorandom sampling strategies for MRI.

Regarding sparseness, an MRI image is often not sparse in the original pixel representation. A subtraction operation can, however, yield a significantly sparser resultant image. In general, mathematical transforms ("sparsifying" transforms) can be applied to a single image to "sparsify" it [113, 114].

For example, the discrete image $f^*(x, y)$ in Figure 5.15(a) can be sparsified by applying a discrete gradient operation, $\Delta f^*(x, y)$, which is defined as

$$\Delta f^*(x, y) = \sqrt{(D_x f^*(x, y))^2 + (D_y f^*(x, y))^2}, \qquad (5.10)$$

where $D_x f^*(x, y) = f^*(x + 1, y) - f^*(x, y)$ and $D_y f^*(x, y) = f^*(x, y + 1) - f^*(x, y)$ to obtain the discrete gradient image shown in Figure 5.15(b); it is significantly sparser than the original. To be more quantitative, by defining a significant pixel to be one with an intensity more than 10% of the highest pixel value in an image, we can use histograms (Figure 5.16) to demonstrate the

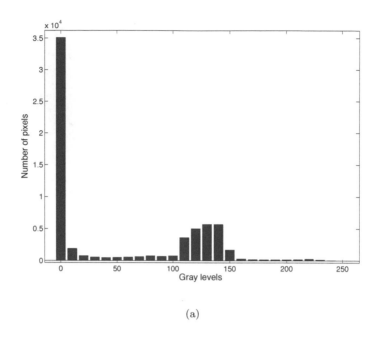

(a)

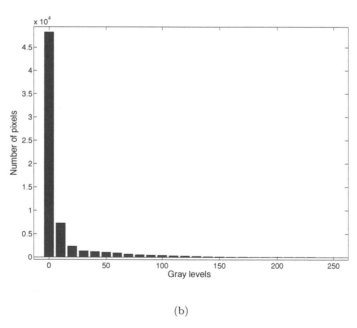

(b)

Figure 5.16
Histograms of (a) the image in Figure 5.15(a) and (b) of its gradient transform Figure 5.15(b).

distribution of the number of significant pixels in the original image and in the discrete gradient image. Based on the ratio between the number of significant pixels in the original and gradient images, the discrete gradient image is 3.5 times sparser than the original image.

To be useful, the sparsification operator must be invertible. That is, from the sparse representation of the image there should be the possibility to return back to the original image. The gradient operator, being invertible, demonstrates that a medical image can be made sparse even if the original is not.

The basic idea in CS image reconstruction theory can be summarized as follows: instead of directly reconstructing a target image, the sparsified version of the image is reconstructed. In the sparsified image, substantially fewer pixels have significant image values. Thus, it is possible to reconstruct the sparsified image from an undersampled data set. After the sparsified image is reconstructed, an "inverse" sparsifying transform is used to reach the target image domain.

To be more specific, let the vector f^* represent an image (an image can be vectorized as a signal). A compression technique compresses the signal by finding some transform Φ (e.g., Fourier transform, wavelet transform, or finite difference as calculated for the image of Figure 5.15 using the gradient operator defined by Equation 5.10) such that $\Phi f^* = g^*$ is approximately sparse. Approximately, sparse implies that some elements of g^* have near-zero contribution to recovering f^*. The next step then is saving the locations and values of those entries of g^* with relatively large magnitudes. To recover f^*, one simply applies the inverse transform $f^* = \Phi^{-1}g^*$. For ease of notation, we set

$s = \|g^*\|_0 :=$ the number of non-zero components[4] in g^*,

$n^2 :=$ dimension$(g^*) =$ dimension(f^*).

The question is: Is it possible to reconstruct f^* uniquely by making $h < n^2$ measurements?

The answer would be no if $s = n^2$. In fact, if a nonzero function, on $s = n^2$ points, g^* exits such that $R^{-1}g^* = f^*$, but g^* has only $h < n^2$ values measured, the remaining $n^2 - h$ values in g^* are arbitrary, and we can define g_1^* such that all these arbitrary values are 1 and g_2^* such that they are 2 (these functions are nonzero on $s = n^2$ points and $g_1^* \neq g_2^*$). We have $R^{-1}g_1^* = f^*$ and $R^{-1}g_2^* = f^*$ implying $R^{-1}(g_1^* - g_2^*) = 0$, that is, $(g_1^* - g_2^*) = 0$. But $(g_1^* - g_2^*) \neq 0$!

This proves that we cannot reconstruct uniquely a function f^* with a number of samples lower than the number of non zero elements. The space of all the images on n^2 points has n^2 degrees of freedom.

If, however, we make the additional hypotheses that $s < n^2$, corresponding to g^* (that is, f^*) having only s degrees of freedom instead of n^2, the answer is yes provided that $h \geq s$, that is, one takes at least as many measurements as the sparsity of f^*.

---

[4]We say that a function f^* is s-sparse if $g^* = \Phi f^*$ is non-zero in at most s places or, equivalently, if its support, that is, the points on which $g^*(x) \neq 0$, has cardinality less than or equal to s.

   This seems reasonable, but a condition must hold; we have to know precisely what the support of f^* is (i.e., we have to know which are the nonzero components). In real situations, we do not have the possibility to know the nonzero components in advance. In this case, the following theorem becomes useful [19]:

Theorem 1 *An s-sparse function f^*, with $2s < n^2$, can be uniquely reconstructed from h samples if $h \geq 2s$.*

Proof 1 *Suppose, for contradiction, that $h < 2s$. Then, the set of functions supported on $\{1, \cdots, 2s\}$ has more degrees of freedom than those measured by the h terms. This means that there exists a function $\hat{f}$ whose $R\hat{f}$ is zero on the collected h terms, but is not identically zero. If we split $\hat{f} = \hat{f}_1 - \hat{f}_2$ where $\hat{f}_1$ is supported on $\{1, \cdots, s\}$ and $\hat{f}_2$ is supported on $\{s + 1, \cdots, 2s\}$, then we say that $\hat{f}_1$ and $\hat{f}_2$ are two distinct s-sparse functions whose transform R agreed on the h terms, thus contradicting the unique recoverability of f^*.*

Ideally, we would like to take the least possible number of measurements, that is, h being equal to s. However, we must pay a price for not knowing the locations of the non-zeros components in g^* (there are n^2 choose s possibilities!) while still asking for perfect reconstructions of most sparse g^*.

   Theorem 1 provides a necessary condition to exactly reconstruct an s-sparse function f^*: the number h of collected measurements b should be at least $2s$. But, is it also a sufficient condition? Or, alternately, how close to $2s$ can be h?

   Some counterexample functions exist [19], indicating that these functions require a number of measurements very close to n^2 in order to be perfectly reconstructed. Most are regular functions, such as the Dirac comb (defined in Chapter 1).

   Random collection of measurements can help against the regular shapes of these counterexample functions with the expectation that an increment of h above $2s$ corresponds an increment of the probability of an exact reconstruction of the unknown f^*. A reasonable number, heuristically estimated, is $h \approx 4s$ (when $4s < n^2$). It remains the case, however, that in MRI we have no way to establish a priori the value of s as the image is unknown.

   The CS process consists, therefore, of three steps: encoding, sensing, and decoding. In the first step, f^* is encoded into a *smaller* vector $b = Rf^*$ of a size $h < n^2$ by a linear transform R. b contains less information than f^*, so it is a *compression* of f^*. Since $f^* = \Phi^{-1}g^*$, $b = Ag^*$ (for $A = R\Phi^{-1}$, $R = A\Phi$) is also a compression of g^*. In many applications of CS, the linear transform R is not calculated by a computer but obtained directly by physical means (for example, MRI collect data directly in the k-space, roughly speaking the Fourier coefficients of the image).

   The second step is obtaining the measurement b from the imaging system. The third step is to recover g^* (and thus f^*) from b. Since g^* is sparse, it is reasonable to seek to recover it as the sparsest solution of the undetermined system of equations $Ag = b$. However, there may exists another *even sparser* solution to these equations, giving rise to the ℓ_0-problem (the ℓ_0 norm of a vector g is simply the number of non-zero elements of g) of finding:

$$\min_g \{\|g\|_0 : Ag = b\}. \tag{5.11}$$

Unfortunately, this problem is intractable for general data, and impractical for nearly all real applications. In fact, this combinatorial optimization problem requires one would seek all subsets of the n^2 elements of cardinality $s \leq \frac{1}{2}h$, checking for each one whether it fills the constrain $Ag = b$. Another possible choice would be to seek the function g of least energy, that is, which minimizes the ℓ_2 norm (the ℓ_2 norm of a vector g is defined as $\|g\|_2 = (\sum_{i=1}^{n^2} |g_i|^2)^{1/2}$):

$$\min_g \{\|g\|_2 : Ag = b\}. \tag{5.12}$$

This method has the advantage of being extremely easy to apply; it is sufficient to consider as zero all the points that have not been measured. Unfortunately, this method does not guarantee the g to be s-sparse. Another computationally tractable alternative is to seek g that minimizes the ℓ_1-norm (defined as $\|g\|_1 = \sum_{i=1}^{n^2} |g_i|$), a process also called *basis pursuit*:

$$\min_g \{\|g\|_1 : Ag = b\}. \tag{5.13}$$

The key difference between this minimization problem and the ℓ_0 problem is that the ℓ_1 norm is convex, so this minimization problem can be solved. Regarding whether the method allows to recover the required s-sparse function, it has been shown [19] that, when we randomly select, using a Gaussian probability distribution function, a set of Fourier coefficients, Equation 5.13 can recover g^* (technically, with a high probability) from b.

More generally, if the function is structured in a perverse way, then the ℓ_1 method will not work. In most practical cases though, the method has been found to yield sparse solutions where the number of collected samples is above $2s$ (see [19] for details) and measurements are performed by using random positions. Moreover, perfect reconstruction is possible from an incomplete dataset if we have the object under reconstruction is piecewise constant [155]. As many objects in MRI applications can be approximately considered as piecewise constant, accurate reconstruction is frequently possible.

The CS image reconstruction selects an image that minimizes the ℓ_1 norm of the sparsified image among all images which are consistent with the physical measurements $Ag = b$. Once g^* is recovered, f^* can be determined through $f^* = \Phi^{-1} g^*$.

The exact recovery of f^* by Equation 5.13 requires that g^* be sparse and $b = Ag^*$, exactly. In practice, especially in imaging, it is often the case that g^* is only approximately sparse; that is, g^* contains a small number of components with magnitudes significantly larger than those of the rest, which are not necessarily zero. Moreover, measurements b are contaminated with noise $\mathbf{n}$, that is, $b = Ag^* + \mathbf{n}$. For both these reasons, an appropriate relaxation of $b = Ag$ should be considered.

The topic of CS stability studies is how accurately a model can recover signals under the above conditions. Stability results established for Equation 5.13 and its extension [18]

$$\min_g \{ \|g\|_1 : \|Ag - b\|_2^2 \leq \sigma^2 \}, \tag{5.14}$$

show that with measurements similar to the exact recovery case, its solution $\widehat{g}$ satisfies

$$\|\widehat{g} - g^*\|_2^2 \leq O(\sigma + s^{-1/2} \|g^* - g^*(s)\|_1),\,^5 \tag{5.15}$$

where g^* is approximately s-sparse and $g^*(s)$ is the s-approximation to g^* by zeroing the $n^2 - s$ least significant components. Clearly, if g^* is exactly s-sparse and σ is set to 0, then Equation 5.15 reduces to the exact recovery of $\widehat{g} = g^*$.

Solving Equation 5.14 is also equivalent to solving the simpler Lagrange relaxation problem:

$$\min_g \mu \|g\|_1 + \frac{1}{2} \|Ag - b\|_2^2. \tag{5.16}$$

This equivalence is in the sense that solving one of the two problems will determine the parameters in the other such that both give the same solution [90]. Given the data A, b, and σ in the Equation 5.14, there exist practical ways (see [43], for example) to estimate an appropriate μ in Equation 5.16.[6]

CS requires the measurement of a relatively small number of "random" linear combinations of the signal values (much smaller than the number of signal samples nominally used in defining it). However, because the underlying signal is compressible, the nominal number of signal samples is a gross overestimate of the "effective" number of "degrees of freedom" of the signal. As a result, the signal can be reconstructed with good accuracy from relatively few measurements by a convex-constrained optimization procedure. The compressed sensing problem for MRI, as discussed above, can be directly derived through the choice of A and b in Equation 5.14; b is the measured k-space data collected by a MRI scanner and $R = A\Phi$ is the FT operator that calculates values exactly on the positions where b are measured. In that setting, CS is claimed to be able to make accurate reconstructions from a small subset of k-space, rather than an entire k-space grid. The original paper by Candes et

---

[5] Let f(x) and g(x) be two functions defined in some subset of the real numbers. One write $f(x) = O(g(x))$ (f is on the order of g) if and only if there exists a position constant M and a real number x_0 such that $|f(x)| < M|g(x)|$ for all $x > x_0$.

[6] Although real medical images are rarely sparse, transformations of the image via an operator can result in sparse images.

al. [19] considered random undersampling of Fourier coefficients, the practical situation of MRI. In order to reconstruct a complete image from the undersampled data, the simpler strategy assumes that the Fourier coefficients at all of the unobserved frequencies are zero (thus reconstructing the image having "minimal energy" under the observation constraints). This method does not perform very well because the reconstructed image has severe nonlocal artifacts caused by zero filling [19]. A good reconstruction algorithm, it seems, would have to guess the values of the missing Fourier coefficients. This in turn is problematic, as predictions of Fourier coefficients from their neighbors are very delicate, due to the global and highly oscillatory nature of the Fourier transform. A more efficient approach is through convex optimization. More details on this very promising method applied to MRI can be found elsewhere [73, 74].

5.3.2.1.1 Compressed Sensing Results

To highlight the effectiveness of CS, we report the results obtained by undersampling a 256×192 MRI image of the brain of a healthy volunteer collected by a GE Signa MRI System at 1.5 T, zero filled to obtain a 256×256 image and used as a test image in Chapter 3 and in Figure 5.15(a).

CS reconstruction has been implemented in Matlab using wavelet transform (Daubechies orthogonal wavelets D4) as sparsifying transform, and ℓ_1-minimization. The k-space representation was undersampled by collecting 100 rows (all the columns were sampled in those rows that were selected), corresponding to 25,600 of 65,536 data, by using a random sampling with a normal distribution (the central, most informative, rows are sampled more densely).

As seen above where the image was used in testing, we achieved a 3.5 fold sparser representation with the gradient operator with respect to the original. This demonstrated the image to be sparse, but it is very difficult to establish the optimal value for s. From the original image, with approximately 33,000 significant pixels (by thresholding to separate tissue and background), we can estimate the value of s is about 9,500 if we consider the gradient operator; the other are occupied by noise and zeroes. It may be that s can be made lower by considering a more efficient sparsifying operator but, as a conservative value due to the presence of noise, 9,500 is a reasonable value for s.

In light of earlier arguments, h should be about 38,000, corresponding to $4s$. For our experiment, we used 25,600 samples, corresponding to about 40% of the whole data set. The results of CS reconstruction using the row data shown in Figure 5.17(b) are shown in Figure 5.17(a) after 9 iterations. For comparison a one-step zero-filled reconstruction, using the same data, is given in Figure 5.17(c). To show the residual aliasing in the CS image, the difference between the complete image and the CS reconstruction is shown in Figure 5.17(d).

Nonlinear reconstruction in CS reduces the number and amplitude of aliasing artifacts that are particularly evident in the linearly reconstructed image

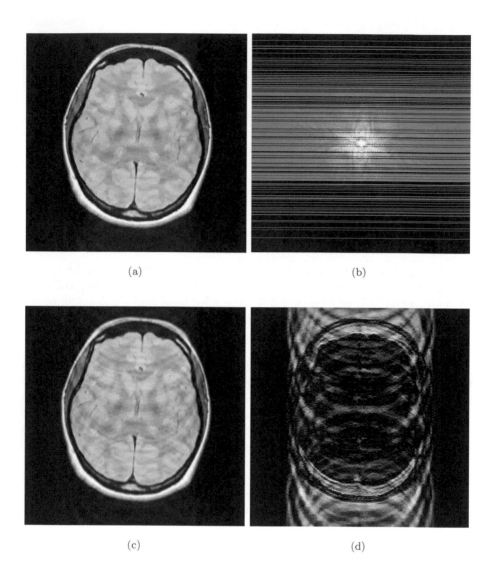

(a)   (b)

(c)   (d)

Figure 5.17
(a) CS results obtained on Cartesian sampled data in which 100 rows have been randomly undersampled by using a Gaussian distribution (25,600 of 65,536 data, about 40% of the total); (b) data rows that have been used for the CS reconstruction; (c) zero-filled image; (d) image obtained as a difference between the completely filled image and CS image. Note gray scale for (d) is extended relative to (a) and (c).

from the same undersampled data set. The artifacts reduction is not, however, particularly evident between Figure 5.17(a) and Figure 5.17(c). This is also reflected in the difference PSNR values, rising from 20.95 dB for the linearly reconstructed image to 23.98 dB for the CS image.

The presence of residual artifacts is due to:

1. A relatively low number of data has been used for reconstruction (about 40% of the total, 67% of the recommended).

2. Parallel line acquisition. In 2D MRI, the only "randomizable" direction is along the phase-encoding gradient, leading to a degree of coherence between the various samples, thus violating the random condition.

5.3.2.1.2 Some Remarks on Compressed Sensing in MRI

MRI, due to its very rich set of imaging parameters, produces images ranging from complete morphological images, such as those reported in Figure 5.9(a) or in Figure 5.15 to functional images or specific images, such as angiography images, that are very different each other. An appropriate sparsifying transform would serve to reduce the number of collected data or to improve the resulting image quality. Due to the diversity of the produced images, the choice of a unique sparsify transformation, which can be efficient for all of them, is currently under investigation.

For example, the total variation norm, based on finite-differencing, can be very useful to sparsify MR angiography images (the resulting sparsified images are sparser than those obtained with other methods), while wavelet transform can be better employed for brain images (the resulting sparsified images are sparser than those obtained with other methods).

Another critical point is the choice of the minimization norm. Though revolutionary, CS can be efficiently applied if the undersampling is not too different from the Nyquist's rate. In fact, most CS applications, especially within medical imaging, have centered on the ℓ_1-minimization approach due to the fact that the corresponding ℓ_0-minimization problem is intractable. Although ℓ_1-based techniques are extremely powerful, they inherently require a degree of sampling above the theoretical minimum sampling rate to guarantee that exact reconstruction can be achieved.

An interesting recent improvement of CS regards the proposal of an innovative Homotopic ℓ_0-minimization [69] for reconstructing MR images at sampling rates below those achievable using ℓ_1-based CS methods by directly attacking the ideal ℓ_0-minimization. The ℓ_0-minimization problem is described along with both its theoretical and applied implications. Most importantly, a practical scheme for addressing the ℓ_0 quasi-norm based on homotopic approximation is presented, and an efficient semi-implicit numerical scheme for computation is described. Both problem tractability and the goodness of results have been demonstrated, when compared to the classical ℓ_1-based CS methods, in spite of the reduction of the samples used for reconstruction.

As reported above, it seems that conventional readout with parallel undersampling of the phase-encode direction produces residual coherent aliasing on CS image reconstruction, as depicted in the reported example. Nevertheless, one can imagine extracting sets of data in a random way from the completely sampled rows, thus resulting in different data sets each resembling a completely random dataset, both along its rows and along its columns. In this case, we would consider the following:

1. Are the extracted data sufficient to obtain feasible CS reconstructions?

2. How will data coherence change, in each of the data sets, with respect to that of the complete rows set?

3. If the obtained coherence is lower than that of the whole data set, can CS reconstructions have reduced coherent aliasing?

4. Can the series of obtained CS reconstructions be recombined to improve SNR without increasing aliasing?

These are questions currently the focus of research. Another way to approach this problem is to pursue a deterministic version of CS which is rigorously backed by theory. This would be very beneficial for imaging purposes, but is yet lacking.

Relating to this last point, the CS technique does not take into account the information content of the image to which it is applied, in contrast to adaptive reconstruction methods.

A step in this direction is to consider whether CS reconstruction can be efficiently applied in an adaptive acquisition context or, in the case that CS is feasible for adaptive acquisition, whether adaptive acquisition methods can be used to improve CS reconstructed images or to reduce the data set without reducing image quality.

As can be noted from the above discussion, CS theory is still in its early stage and it is currently in refinement along different lines. In what follows we show some preliminary results that can help to answer the last point, that is, to apply CS on adaptively collected sampled.

5.3.2.2 Compressed Sensing on Adaptively Collected Data

Basis for believing the CS strategy can in fact be applied to adaptively collected data arises from prior reports of reconstruction from perturbed projections (Figure 5.14d), because those data have a high level of incoherence (very similar level to what can be obtained by variable density perturbed spirals of Figure 5.14f).

Further, as noted for the CS example considered above, the image quality would benefit from sampling densely that correspondences to the energy distribution in k-space. But this is what adaptive acquisition methods do!

To put these ideas in practice, we compare, using the same test image, the traditional CS strategy applied on randomly collected data sets, using both uniform random undersampling and a 2D Gaussian distribution to weight the samples to the central k-space region (reflecting the fact that the majority of image energy is contained near the center of the k-space) with a CS strategy applied on adaptively sampled data sets. The adaptive trajectories considered include both radial projections (using the method described above) and randomly sampled data collected in square sub-windows of k-space that adapt to where the information content is maximum.

The sub-windows method is a particular case of adaptive acquisition with the variation that the k-space samples can be randomly collected inside squares.

To describe the data in square sub-windows, information content defined as

$$I = \frac{E \cdot SquareArea}{DataNumber} \tag{5.17}$$

is used in the iterative algorithm 5.3. In this definition, E represents the entropy function defined in Equation 5.2, $SquareArea$ is the area of the current square (this term serves to stimulate acquisition on big squares) and $DataNumber$ is the number of k-space data present in the given square (it serves to normalize the information content to the mean value in the given square). Algorithm 5.3 seeks to collect the first $MaxNumber$ data by fol-

---

Algorithm 5.3: Calculate the, approximately, most informative k-space set.

Assign $CurrentSquare$ = whole k-space set;
Assign $Number = 0$;
while $Number < MaxNumber$ **do** {The adaptive acquisition
terminates when the desired number of coefficients has been reached.}
    Uniform random extraction of $MIN(20 \%$ of the $CurrentSquare$ data
    set, $MaxNumber - Number$) values;
    Upgrade $Number$;
    Divide $CurrentSquare$ in four quadrants;
    Calculate I for the resultant squares;
    Upgrade $CurrentSquare$ to the square whose I is maximum AND
    square side is ≥ 16 pixels;
end while
Return the measured data set;

---

lowing a trajectory defined by subdividing squares in k-space on the basis of maximum mean information, excluding square of side lower than 16 pixels. A typical sampling pattern for the algorithm is illustrated in Figure 5.18. During its execution, the algorithm collects first data from the whole square, then converges on internal smaller squares, and finally refines its search on external bigger squares.

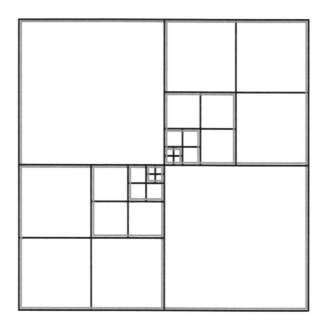

Figure 5.18
Subwindows adaptive acquisition scheme: an example.

Numerical experiments were performed using this algorithm, as an example, to allow comparison of standard randomly measured data with a data set obtained following, in an adaptive way, the image energy distributed among k-space zones, not necessarily represented by complete radial directions. This last method can be considered a form of intelligent random acquisition; incoherence can be maintained and information content can be introduced.

Simulations with different numbers of data samples were performed to highlight the behavior of different acquisition techniques on the reconstruction results. For each reconstruction, we calculated the PSNR, as summarized in Figure 5.19.

The results obtained with the CS combined with each of the four acquisition techniques are shown in Figures 5.20–5.23 for the case of 25,600 samples, about 40% of the total. In particular, Figure 5.20 shows the CS when applied to uniformly random undersampling; Figure 5.21 shows the CS when applied to random undersampling weighted with a Gaussian function; Figure 5.22 shows the CS when applied to radial adaptively collected projections; and

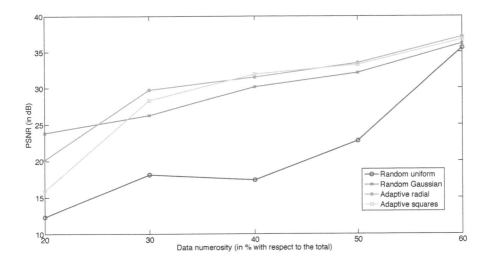

Figure 5.19
PSNR values of the CS reconstructions by different acquisition modalities. The data set numerosity ranged from 20% of the total to 60% of the total in steps of 10%.

Figure 5.23 shows the CS when applied to adaptively collected square subwindows.

These results can also be compared with those in Figure 5.17 where CS was applied on parallel undersampling with the same number of samples, far from the estimated necessary 38,000 samples.

As can be observed from Figure 5.19, and confirmed by visual comparison of the reconstructed images, with 40% of the data, uniform random sampling generally performed worse than the other methods. The reason is that most of the used data sets had a number of samples which is far from the estimated $4s$ (except the last, for which the number of samples is above the estimated gold number, 38,000), but also because the sampling does not provide an adequate sparse representation of the data (a random basis is not necessarily a sparse basis).

The other acquisition techniques performed approximately in the same in terms of PSNR, and the results were all comparable with the original both in aliasing reduction and in resolution.

Uniform random sampling PSNR curve had an increment which is faster than the other methods; this is also justified by the fact that all the methods converge to the optimal result as the number of samples rises. Moreover, it shows a fluctuating trend between 30% and 40% of data; this is likely because the dataset is too incomplete and reconstruction is not stable (the dominant energy features not having necessarily yet been captured). The behavior of CS on randomly chosen samples confirmed both that the estimated value for s was not so far from its true value and that CS performs better if applied on data having greater amplitude, that is higher energy (that is carrying more information).

Gaussian random sampling, due to the fact that it sampled the k-space center most densely, started with greater PSNR values than the other methods. Nevertheless, as the number of samples increased from 20% to 30%, it was overcome by the adaptive techniques. As can be noted also from the visual comparison, Gaussian random sampling failed to describe the edges, corresponding to a considerable loss of spatial resolution. This persisted for the data sets with greater numbers of samples. It is recognized that PSNR is not a good parameter to evaluate resolution maintenance.

It is worth noting that PSNR for both adaptive methods were worse than Gaussian random sampling when just 20% of samples were used, due to the fact that these methods orientated both in low-frequency and high-frequency acquisition. Moreover, the adaptive square method, because of the forced sampling structure, does not densely sample any region in the early stages of acquisition and, in consequence, some acquisition steps are necessary to adapt toward maximum energy data. In this case, it was sufficient to reach 30% of the data to outperform the Gaussian sampling. The goodness of their results, both in artifacts reduction and resolution maintenance, is also demonstrated by visual comparison.

Both radial and square adaptive methods give very similar results where more data was used; with 40% of the data, they gave results comparable with those the other methods achieved with 60% of the data.

As last note, the CS adaptive methods performed better than simple CS applied to Cartesian sampled data, especially regarding artifacts reduction; in fact, the simple CS result reported in Figure 5.17 had a PSNR value of 23.98, obtained with 40% of the data, which was lower than the PSNR obtained by both adaptive CS methods with 30% of the data. In conclusion, we can affirm that the adaptively collected data are consistent with CS reconstruction. Moreover, they give the possibility to go below the data set dimensions necessary for simple random sampling methods to obtain good reconstruction results. This also implies that adaptive methods can be efficiently used to reduce acquisition time. This is very promising, though preliminary, and has to be confirmed through the application on other images and deserves further exploration in future.

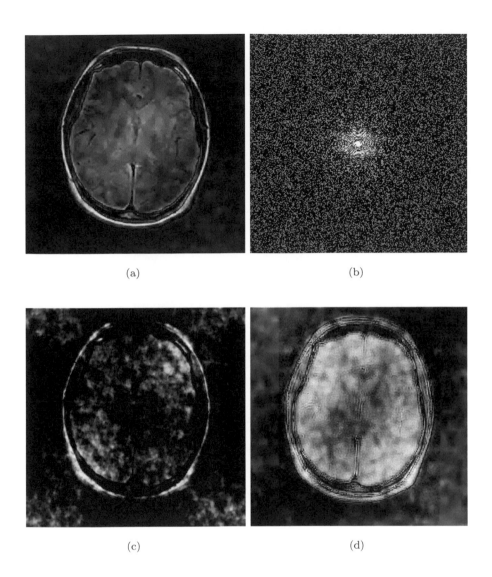

(a) (b)

(c) (d)

Figure 5.20
(a) CS results obtained on randomly undersampled data (25,600 of 65,536 data, about 40% of the complete set); (b) data rows that have been used for the CS reconstruction; (c) zero-filled image; (d) image obtained as a difference between the completely filled image and CS image. Note gray scale for (d) is extended relative to (a) and (c).

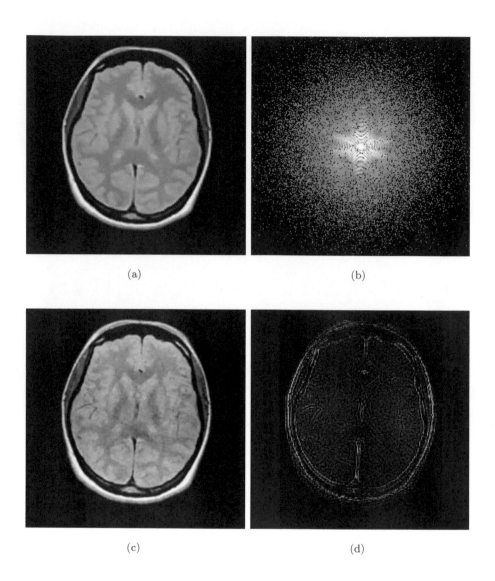

(a)   (b)

(c)   (d)

Figure 5.21

(a) CS results obtained on heavily Gaussian undersampled data (25,600 of 65,536 data, about 40% of the complete set); (b) data rows that have been used for the CS reconstruction; (c) zero-filled image; (d) image obtained as a difference between the completely filled image and CS image. Note gray scale for (d) is extended relative to (a) and (c).

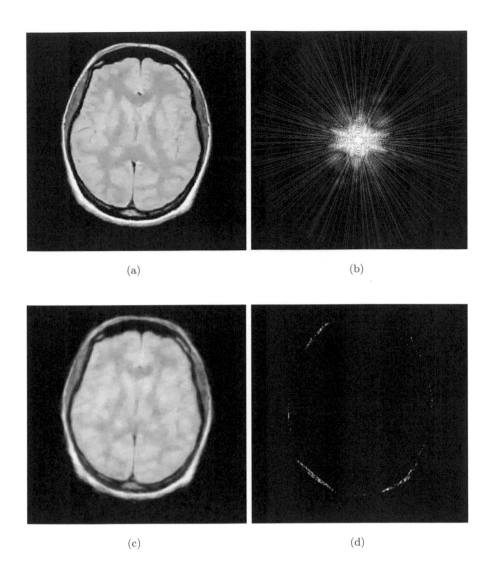

(a)

(b)

(c)

(d)

Figure 5.22
(a) CS results obtained on undersampled radially collected adaptive data
(25,600 of 65,536 data, about 40% of the complete set); (b) projections that
have been used for the CS reconstruction; (c) zero-filled image; (d) image obtained as a difference between the completely filled image and CS image. Note
gray scale for (d) is extended relative to (a) and (c).

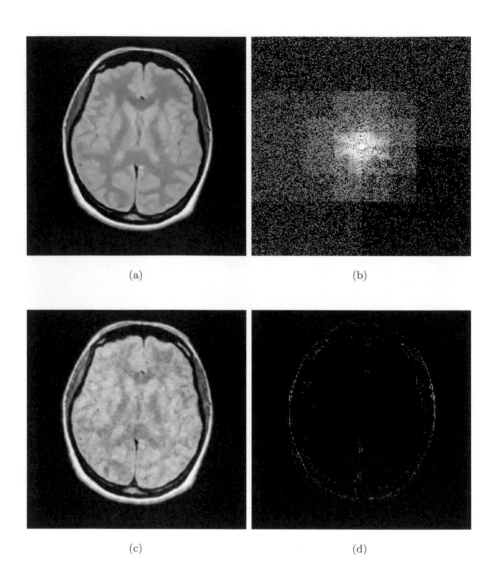

(a)   (b)

(c)   (d)

Figure 5.23
(a) CS results obtained on undersampled data adaptively collected on the most
informative subwindows (25,600 of 65,536 data, about 40% of the complete
set); (b) data that have been used for the CS reconstruction; (c) zero-filled
image; (d) image obtained as a difference between the completely filled image
and CS image. Note gray scale for (d) is extended relative to (a) and (c).

5.4 Summary

Recently, MRI has developed considerably in the directions of dynamic and functional imaging, opening up several new fields of application, in particular, real-time imaging. Although these developments are generally promising, their application can be limited by the compromise between temporal and spatial resolution. In order to improve temporal resolution, many have turned to undersampling. Some of the most effective methods for sparse sampling acquisition and reconstruction, with the aim of reducing the undersampling artifacts, have been reviewed with an eye to their membership in one of two classes.

The first class covers those acquisition/reconstruction/restoration methods that involve adapting the trajectories during the acquisition (i.e., the chosen trajectories, both in number and directions, are dependent on the sample shape), while the second includes reconstruction/restoration methods that reduce artifacts independently of the sample shape. The first class contains some adaptive methods that allow a substantial reduction of the collected data if the k-space paths are collected through the most informative locations (coefficients with maximum energy). To the second class belongs compressed sensing, a breakthrough technique which, to some extent, appears counterintuitive; images can be almost exactly reconstructed from a sparse set of their Fourier coefficients, if they are compressible and the collected coefficients are randomly placed in k-space.

Both adaptive methods and CS require the images to be compressible, but they are completely different regarding the priorities determining where to sample; in the first data are collected along the "most informative" trajectories (radial directions or squares of different side), where energy is maximum, in the second data are collected randomly.

In principle, both strategies appear reasonable and a number of reports have shown promising results for each in one or another application, but some questions have arisen. Must these strategies be considered separate and, apparently, in contrast each other, or is it possible to reach a point of convergence between smart acquisition and randomly collected data?

To this end, we presented preliminary results demonstrating that CS hypotheses can be filled with data collected by adaptive methods and, in consequence, convex reconstruction can be used on adaptively collected CS data. The results obtained demonstrated that CS on adaptively collected data can yield good reconstruction with less data than classical random CS. This could represent another starting point for an exciting research for the future of MRI.

Part IV

The Future

6

Conclusions and perspectives

CONTENTS

Some great challenges still attend MRI, if it is to meet the expectations of the growing numbers of potential applications. Functional and molecular imaging, for example, require that MRI be capable of ever more highly temporally and spatially resolved images, with high contrast between different tissues, high sensitivity, and high SNR. This has driven a trend in the direction of very high field superconductive magnets (7 T or greater) to improve sensitivity and, consequently, temporal resolution. At the same time, the magnetic field has to be better shimmed to also improve spatial resolution while avoiding distortions, fundamental if MRI has to be used as a precise localizing technique. Similar demands and responses are to be seen in real-time imaging.

Against this tide, there is the desire to make the magnet more comfortable: for the patients, to avoid claustrophobia and constrained positions and, for the physicians, to allow better access for therapy and choice of patient position to allow MRI to be used for interventional medical procedures.

A further challenge is to produce MRI equipments that efficiently, ideally simultaneously, collect images from different body districts, for example to perform whole-body imaging at a cost that would allow the mass screening of the whole population. Current whole-body MRI scanners are too narrow for very heavy people, and are not in fact whole-body scanners but rather move the patient along the magnet to image the body (in steps of about 45 cm). The direction, in particular, is at odds with needs for high homogeneity in ever stronger magnets.

A very desirable concept for a scanner would be a completely flat, open magnet, extending from the head to the feet of the person, like a bed. To the best of our knowledge, however, the magnetic field produced by such a geometry should be too inhomogeneous to allow conventional imaging. To this point, we have tried in this book to give a different answer, showing that, in principle, it could be possible to perform imaging, without distortions, from residual magnetic field inhomogeneity which is greater than that acceptable for presently used acquisition sequences. There remains considerable experimental activity to evaluate the potentialities and the limitations of the method proposed in Chapter 4 of this book. To complete the discussion on this point, a question remains: would a completely open whole-body MRI scanner be cheaper than existing designs? Two encouraging factors are that the expensive shimming procedure could be avoided, and the duration of an examination could be shortened.

Second crucial point examined here is that of reducing imaging time. Along this direction, MRI is undergoing a revolution: with a proper choice of smart coding methods and acquisition sequences, either using a priori information about the sample or by using random sparse sampling and reconstruction (through compressed sensing) MRI can make substantial advances with little hardware modification.

Adaptive techniques for acquisition from projections were proposed by the author and colleagues as early as 1996 in the field of Electron Paramagnetic Resonance Imaging [103], where acquisition time reduction is very crucial (the time necessary to collect a single projection can be of tens of seconds in continuous wave imaging and, for this reason, a complete image can require several minutes). These methods were extended to nuclear MRI in 2000 [105]. The important innovation regarded two aspects: for the first time, the possibility to go below the Nyquist's limits was introduced, by using the information about the shape/symmetry/sparsity of the sample during the progress of acquisition procedure (during data collection, information was extracted from the available data and successive directions were collected where the information was maximum); a method to evaluate where the significant data are located was introduced.

In 2005 [18], compressed sensing formalized mathematically that it can be possible to go below the Nyquist's limits and that an image can be reconstructed by using a convex optimization strategy, with no artifacts, if some assumptions are met (image sparsity in some basis, data uncoherence, convex reconstruction). This method was introduced for the first time and tested for MRI in 2007 [73]. Due to the facts that the degree of image sparsity is unknown in advance, and that distribution of this sparsity is also unknown, practical application requires a sort of oversampling above the image sparsity (4–5 times the image sparsity). The extreme sparsity of MR images however means that even with this oversampling above sparsity, compressed sensing continues to be an undersampling regarding the Nyquist's limits (regarding image dimensions).

The present book sought, for the first time, to compare adaptive methods with compressed sensing and demonstrated for the first time that it possible to join the use of adaptive methods with those of compressed sensing.

In Chapter 5 of this book, very preliminary numerical results, showed that:

1. data collected with an adaptive method can be sufficiently noncoherent to allow use of convex reconstruction to obtain an image where undersampling artifacts are almost absent;

2. for a given quantity of sample data, the quality with adaptive compressed sensing can be higher than classical compressed sensing be-

cause an adaptive method allows to collect information both on image *sparsity* and on *sparsity* locations (information about the locations of significant data is collected during acquisition).

This last point is crucial; an adaptive method helps compressed sensing both to establish what the s-sparsity value of an unknown image is, and at the same time, to collect information about significant data position.

It follows, in principle, that it may become possible to reconstruct exactly an image, without undersampling artifacts, just by collecting the minimum p significant data; the reconstruction process can be limited in solving a linear system of p equations in p unknown. The themes treated in this book, for a good part in a speculative sense with preliminary results obtained on synthetic data, deserve to be explored and completely demonstrated with experimental and repeated measurements.

6.1 Some Hypothesis

In summary, the ambition of some is for MRI to overcome its closed imaging systems. C-shaped magnets are on the market, with bores greatly more open than the first cylindrical systems, but this is still insufficient to allow many innovative applications for MRI. The ideal is to realize completely open, flat, really whole body, independent of residual magnetic field inhomogeneity, equipments to be used for imaging during surgical interventions, as a whole-body massive imaging system to prevent tumors (or to discover them in the early stage) or for other, actually still inconceivable, applications. The establishment of methods that allow imaging in inhomogeneous field, as introduced here, removes a significant obstacle to the use of such equipments.

The proposed innovative coding/decoding strategy based on acceleration could be used to define an innovative rich set of acquisition sequences or to change/integrate the actually existing methods. Moreover, it can contribute to improve the fastness of acquisition; a signal could be collected in few hundreds of microseconds instead of some milliseconds. The rest of time could be spent either to improve SNR or to collect new directions, thus improving temporal resolution.

In the same direction moves the proposal of innovative acquisition/reconstruction strategies: to reconstruct "perfect" image by using a reduced data set. In this case, the temporal improvement can be also of a factor of 4 with respect to standard complete imaging.

At its birth, about thirty years ago, scientists hoped that it would become an imaging modality whose images had the same quality as those of CT; no one imagined how the MRI would have been now. In the same way, we cannot imagine now what MRI may become and how it might be transformed. To perform imaging without a magnet, on the earth magnetic field, was inconceivable just a few years ago, but is now an area of active research. The themes treated in this book could represent some of the exciting starting points for a future intriguing evolution of MRI.

Bibliography

[1] K. Arfanakis, A.A. Tamhane, J.G. Pipe, and M.A. Anastasio. k-space undersampling in propeller imaging. *Magn Reson Med*, 53(3):675–83, Mar 2005.

[2] D. Atkinson, D.L. Hill, P.N. Stoyle, P.E. Summers, S. Clare, R. Bowtell, and S.F. Keevil. Automatic compensation of motion artifacts in mri. *Magn Reson Med*, 41(1):163–70, Jan 1999.

[3] D. Atkinson and D.L.G. Hill. Reconstruction after rotational motion. *Magnetic Resonance In Medicine*, 49(1):183–187, January 2003.

[4] D. Atkinson, D.A. Porter, D.L. Hill, F. Calamante, and A. Connelly. Sampling and reconstruction effects due to motion in diffusion-weighted interleaved echo planar imaging. *Magn Reson Med*, 44(1):101–9, Jul 2000.

[5] E.E. Babcock, L. Brateman, J.C. Weinreb, S.D. Horner, and R.L. Nunnally. Edge artifacts in MR images: chemical shift effect. *Journal of computer assisted tomography*, 9(2):252, 1985.

[6] A.V. Barger, W.F. Block, Y. Toropov, T.M. Grist, and C.A. Mistretta. Time-resolved contrast-enhanced imaging with isotropic resolution and broad coverage using an undersampled 3d projection trajectory. *Magn Reson Med*, 48(2):297–305, Aug 2002.

[7] P. Barone and R. March. On the super-resolution properties of prony's method. *Zeitschrift Fur Angewandte Mathematik Und Mechanik*, 76:177–180, 1996.

[8] N. Ben-Eliezer, Y. Shrot, and L. Frydman. High-definition, single-scan 2d mri in inhomogeneous fields using spatial encoding methods. *Magn Reson Imaging*, 28(1):77–86, Jan 2010.

[9] P. Bendel and L.R. Zion. Method to eliminate the effects of magnetic field inhomogeneities in nmr imaging and apparatus therefor. *U.S. Patent N.US4656425*, 1987.

[10] M.A. Bernstein, K.E. King, X.J. Zhou, and W. Fong. *Handbook of MRI pulse sequences*, volume 32. 2005.

[11] F. Bloch, W.W. Hansen, and M.D. Packard. Nuclear introduction. *Phys Rev*, 69:127–132, 1946.

[12] R.N. Bracewell. *The fourier transform & its applications 3rd Ed.* 2000.

[13] E.O. Brigham. *The Fast Fourier Transform and its applications*. Signal Processing. Prentice Hall, 1988.

[14] R.A. Brooks and G. Di Chiro. Principles of computer assisted tomography (cat) in radiographic and radioisotopic imaging. *Phys Med Biol*, 21(5):689–732, Sep 1976.

[15] G.G. Brown, J.E. Perthen, T.T. Liu, and R.B. Buxton. A primer on functional magnetic resonance imaging. *Neuropsychology Review*, 17(2):107–125, 2007.

[16] M.E. Brummer, D. Moratal-Pérez, C.Y. Hong, R.I. Pettigrew, J. Millet-Roig, and W.T. Dixon. Noquist: Reduced field-of-view imaging by direct fourier inversion. *Magnetic Resonance in Medicine*, 51(2):331–342, 2004.

[17] A. Buecker, J.M. Neuerburg, G.B. Adam, C.C. Nolte-Ernsting, D.W. Hunter, A. Glowinski, J.J. van Vaals, and R.W. Guenther. Mr-guided percutaneous drainage of abdominal fluid collections in combination with x-ray fluoroscopy: initial clinical experience. *Eur Radiol*, 11(4):670–4, 2001.

[18] E.J. Candes, J. Romberg, and T. Tao. Stable Signal Recovery from Incomplete and Inaccurate Measurements. *ArXiv Mathematics e-prints*, March 2005.

[19] E.J. Candes, J. Romberg, and T. Tao. Robust uncertainty principles: Exact signal reconstruction from highly incomplete frequency information. *Ieee Transactions On Information Theory*, 52(2):489–509, February 2006.

[20] E.J. Candès and T. Tao. Near-optimal signal recovery from random projections: Universal encoding strategies? *IEEE Transactions on Information Theory*, 52(12):5406–5425, 2006.

[21] J. Chen, L. Zhang, J. Luo, and Y. Zhu. Mri reconstruction from 2d partial k-space using pocs algorithm. In *Bioinformatics and Biomedical Engineering , 2009. ICBBE 2009. 3rd International Conference on*, pages 1 –4, june 2009.

[22] S. Clasen and P.L. Pereira. Magnetic resonance guidance for radiofrequency ablation of liver tumors. *J Magn Reson Imaging*, 27(2):421–33, Feb 2008.

[23] I. Contreras, A. Guesalga, M. P. Fernandez, M. Guarini, and P. Irarrazaval. Mri fast tree log scanning with helical undersampled projection acquisitions. *Magn Reson Imaging*, 20(10):781–7, Dec 2002.

[24] W.C. Crowley and R.H. Freeman. Remotely positioned mri system. *U.S. Patent N.US5304930*, 1994.

[25] W.C. Crowley and R.H. Freeman. Method for maintaining encoded coherence for remotely positioned mri device. *U.S. Patent N.US5493225*, 1996.

[26] R. Damadian. Tumor detection by nuclear magnetic resonance. *Science*, 171(976):1151–3, Mar 1971.

[27] R. Damadian. Apparatus and method for detecting cancer in tissue. *U.S. Patent N.US3789832*, 1974.

[28] S.R. Deans. *The Radon transform and some of its applications*. Wiley New York, 1983.

[29] D.L. Donoho. Compressed Sensing. *IEEE Transactions on Information Theory*, 52(4):1289–1306, 2006.

[30] J.M.S. Edelstein, W.A. Hutchison, G. Johnson, and T.W. Redpath. Spin warp NMR imaging and applications to human whole-body imaging. *Physics in Medicine and Biology*, 25:751, 1980.

[31] C.L. Epstein. Magnetic resonance imaging in inhomogeneous fields. *Inverse Problems*, 20(3):753–780, jun 2004.

[32] C.L. Epstein and J. Magland. A novel technique for imaging with inhomogeneous fields. *J Magn Reson*, 183(2):183–92, Dec 2006.

[33] C.L. Epstein and J. Magland. Methods and apparatus for magnetic resonance imaging in inhomogeneous fields. *U.S. Patent N.US7309986*, 2007.

[34] D.A. Feinberg, J.D. Hale, J.C. Watts, L. Kaufman, and A. Mark. Halving mr imaging time by conjugation: demonstration at 3.5 kg. *Radiology*, 161(2):527–531, 1986.

[35] J.H. Gao, J. Xiong, S. Lai, E.M. Haacke, M.G. Woldorff, J. Li, and P.T. Fox. Improving the temporal resolution of functional mr imaging using keyhole techniques. *Magn Reson Med*, 35(6):854–60, Jun 1996.

[36] A. Glowinski, G. Adam, A. Bücker, J. Neuerburg, J.J. van Vaals, and R.W. Günther. Catheter visualization using locally induced, actively controlled field inhomogeneities. *Magn Reson Med*, 38(2):253–8, Aug 1997.

[37] R.C. Gonzalez and R.E. Woods. *Digital image processing.* Prentice Hall, 2008.

[38] M.A. Griswold, P.M. Jakob, R.M. Heidemann, M. Nittka, V. Jellus, J. Wang, B. Kiefer, and A. Haase. Generalized autocalibrating partially parallel acquisitions (grappa). *Magn Reson Med*, 47(6):1202–10, Jun 2002.

[39] H. Gudbjartsson and S. Patz. The rician distribution of noisy mri data. *Magn Reson Med*, 34(6):910–4, Dec 1995.

[40] E.M. Haacke, R.W. Brown, M.R. Thompson, and R. Venkatesan. *Magnetic Resonance Imaging: Physical Principles and Sequence Design.* Wiley-Liss, 1999.

[41] E.M. Haacke, E.D. Lindskog, and W. Lin. A fast, iterative, partial-fourier technique capable of local phase recovary. *Journal of Magnetic Resonance*, 92(1):126–145, March 1991.

[42] E.L. Hahn. Nuclear induction due to free Larmor precession. *Physical Review*, 77(2):297–298, 1950.

[43] E. Hale, W. Yin, and Y. Zhang. Fpc: A fixed-point continuation method for l1-regularization. Technical report, http://www.caam.rice.edu/ optimization/, 2007.

[44] M.S. Hansen, C. Baltes, J. Tsao, S. Kozerke, K.P. Pruessmann, and H. Eggers. k-t blast reconstruction from non-cartesian k-t space sampling. *Magn Reson Med*, 55(1):85–91, Jan 2006.

[45] P.R. Harvey. Real time magnetic field mapping using mri. *U.S. Patent N.US6275038*, 2001.

[46] J. Hennig, A. Nauerth, and H. Friedburg. Rare imaging: a fast imaging method for clinical mr. *Magn Reson Med*, 3(6):823–33, Dec 1986.

[47] K. Homma, N. Nitta, T. Numano, T. Nakatani, K. Hyodo, L. Li, and K. Hikishima. A new half-fourier image reconstruction for mri. In R. Magjarevic, J. H. Nagel, and Ratko Magjarevic, editors, *World Congress on Medical Physics and Biomedical Engineering 2006*, volume 14 of *IFMBE Proceedings*, pages 1583–1586. Springer Berlin Heidelberg, 2007.

[48] J.P. Hornack. *The basics of MRI.* Retrieved from http://www. cis. rit. edu/htbooks/mri/inside. htm, 2002.

[49] D.I. Hoult. Rotating frame zeugmatography. *Journal of Magnetic Resonance (1969)*, 33(1):183 – 197, 1979.

[50] D.I. Hoult. Rotating frame zeugmatography. *Phil. Trans. Royal Soc. of London. Series B, Biol. Sciences*, 289(1037):543547, 1980.

[51] X. Hu and T. Parrish. Reduction of field of view for dynamic imaging. *Magn Reson Med*, 31(6):691–4, Jun 1994.

[52] Y. Hu and G.H. Glover. Three-dimensional spiral technique for high-resolution functional mri. *Magn Reson Med*, 58(5):947–51, Nov 2007.

[53] Y. Hu and G.H. Glover. Increasing spatial coverage for high-resolution functional mri. *Magn Reson Med*, 61(3):716–22, Mar 2009.

[54] P. Irarrázaval, J.M. Santos, M. Guarini, and D. Nishimura. Flow properties of fast three-dimensional sequences for mr angiography. *Magn Reson Imaging*, 17(10):1469–79, Dec 1999.

[55] G.M. Israel, M. Korobkin, C. Wang, E.N. Hecht, and G.A. Krinsky. Comparison of unenhanced CT and chemical shift MRI in evaluating lipid-rich adrenal adenomas. *American Journal of Roentgenology*, 183(1):215, 2004.

[56] D.L. Janzen, D.G. Connell, A.L. MacKay, and Q.S. Xiang. Method of correcting for magnetic field inhomogeneity in magnetic resonance imaging. *U.S. Patent N.US6150815*, 2000.

[57] P. Jezzard and R.S. Balaban. Correction for geometric distortion in echo planar images from b0 field variations. *Magn Reson Med*, 34(1):65–73, Jul 1995.

[58] P. Jezzard and S. Clare. Sources of distortion in functional MRI data. *Human Brain Mapping*, 8(2-3):80–85, 1999.

[59] R.A. Jones, O. Haraldseth, T.B. Müller, P.A. Rinck, and A.N. Oksendal. K-space substitution: a novel dynamic imaging technique. *Magn Reson Med*, 29(6):830–4, Jun 1993.

[60] Y.M. Kadah and X. Hu. Algebraic reconstruction for magnetic resonance imaging under B0 inhomogeneity. *Medical Imaging, IEEE Transactions on*, 17(3):362–370, 2002.

[61] L. Kaufman, J.W. Carlson, D.M. Kramer, J.D. Hale, and K. Yee. Method and apparatus for compensating magnetic field inhomogeneity artifact in mri. *U.S. Patent N.US5157330*, 1992.

[62] L. Kaufman, L.E. Crooks, D.M. Kramer, K. Hake, H. Avram, and J. Wummer. Mri compensated for spurious rapid variations in static magnetic field during a single mri sequence. *U.S. Patent N.US4970457*, 1990.

[63] A.B. Kerr, J.M. Pauly, B.S. Hu, K.C. Li, C.J. Hardy, C.H. Meyer, A. Macovski, and D.G. Nishimura. Real-time interactive mri on a conventional scanner. *Magn Reson Med*, 38(3):355–67, Sep 1997.

[64] A.M. Kinsey, C.J. Diederich, V. Rieke, W.H. Nau, K. Butts Pauly, D. Bouley, and G. Sommer. Transurethral ultrasound applicators with dynamic multi-sector control for prostate thermal therapy: in vivo evaluation under mr guidance. *Med Phys*, 35(5):2081–93, May 2008.

[65] K.K. Kwong, J.W. Belliveau, D.A. Chesler, I.E. Goldberg, R.M. Weisskoff, B.P. Poncelet, D.N. Kennedy, B.E. Hoppel, M.S. Cohen, and R. Turner. Dynamic magnetic resonance imaging of human brain activity during primary sensory stimulation. *Proc Natl Acad Sci U S A*, 89(12):5675–9, Jun 1992.

[66] P.C. Lauterbur. Image formation by induced local interactions: examples employing nuclear magnetic resonance. *Nature*, 242(5394):190–191, 1973.

[67] P.C. Lauterbur. Magnetic resonance zeugmatography. *Pure Appl. Chem.*, 40(1-2):149–157, 1974.

[68] J.H. Lee, B.A. Hargreaves, B.S. Hu, and D.G. Nishimura. Fast 3d imaging using variable-density spiral trajectories with applications to limb perfusion. *Magn Reson Med*, 50(6):1276–85, Dec 2003.

[69] D.A. Leung, J.F. Debatin, S. Wildermuth, N. Heske, C.L. Dumoulin, R.D. Darrow, M. Hauser, C.P. Davis, and G.K. von Schulthess. Real-time biplanar needle tracking for interventional mr imaging procedures. *Radiology*, 197(2):485–8, Nov 1995.

[70] J.R. Liao, J.M. Pauly, T.J. Brosnan, and N.J. Pelc. Reduction of motion artifacts in cine mri using variable-density spiral trajectories. *Magn Reson Med*, 37(4):569–75, Apr 1997.

[71] S. Ljunggren. A simple graphical representation of Fourier-based imaging methods. *J. Magn. Reson*, 54(2):338–343, 1983.

[72] R. Lufkin, L. Teresi, L. Chiu, and W. Hanafee. A technique for mr-guided needle placement. *AJR Am J Roentgenol*, 151(1):193–6, Jul 1988.

[73] M. Lustig, D. Donoho, and J.M. Pauly. Sparse mri: The application of compressed sensing for rapid mr imaging. *Magn Reson Med*, 58(6):1182–95, Dec 2007.

[74] M. Lustig, D.L. Donoho, J.M. Santos, and J.M. Pauly. Compressed sensing mri. *Signal Processing Magazine, IEEE*, 25(2):72 –82, march 2008.

[75] J.R. MacFall, N.J. Pelc, and R.M. Vavrek. Correction of spatially dependent phase shifts for partial fourier imaging. *Magnetic Resonance Imaging*, 6(2):143 – 155, 1988.

[76] B. Madore, G.H. Glover, and N.J. Pelc. Unaliasing by fourier-encoding the overlaps using the temporal dimension (unfold), applied to cardiac imaging and fmri. *Magn Reson Med*, 42(5):813–28, Nov 1999.

[77] P. Mansfield. Multi-planar image formation using nmr spin echoes. *J. Phys. C: Solid State Phys.*, 10(3):L55–L58, 1977.

[78] P. Mansfield. Real-time echo-planar imaging bv nmr. *Br Med Bull*, 40:187–190, 1984.

[79] G. Margosian, P. M. DeMeester and H. Liu. Partial fourier acquisition in mri. *Encyclopedia of Magnetic Resonance.*, 2007.

[80] G.J. Marseille, R. de Beer, M. Fuderer, A.F. Mehlkopf, and van Ormondt D. Nonuniform phase-encode distributions for mri scan time reduction. *J Magn Reson B*, 111(1):70–5, Apr 1996.

[81] R. Matsumoto, A.M. Selig, V.M. Colucci, and F.A. Jolesz. Mr monitoring during cryotherapy in the liver: predictability of histologic outcome. *J Magn Reson Imaging*, 3(5):770–6, 1993.

[82] K. McLeish, D.L.G. Hill, D. Atkinson, J.M. Blackall, and R. Razavi. A study of the motion and deformation of the heart due to respiration. *IEEE Trans Med Imaging*, 21(9):1142–50, Sep 2002.

[83] R.M. Mersereau and A.V. Oppenheim. Digital Reconstruction of Multidimensional Signals from Their Projections, proc. of IEEE, vol. 62. *No*, 10:1319–1338, 1974.

[84] C.H. Meyer, B.S. Hu, D.G. Nishimura, and A. Macovski. Fast spiral coronary artery imaging. *Magn Reson Med*, 28(2):202–13, Dec 1992.

[85] C.H. Meyer and P. Irarrazabal. Magnetic field inhomogeneity correction in mri using estimated linear magnetic field map. *U.S. Patent N.US5617028*, 1997.

[86] R. Mir, A. Guesalaga, J. Spiniak, M. Guarini, and P. Irarrazaval. Fast three-dimensional k-space trajectory design using missile guidance ideas. *Magn Reson Med*, 52(2):329–36, Aug 2004.

[87] C.A. Mistretta, O. Wieben, J. Velikina, W. Block, J Perry, Y. Wu, K. Johnson, and Y. Wu. Highly constrained backprojection for time-resolved mri. *Magn Reson Med*, 55(1):30–40, Jan 2006.

[88] D.D. Muresan and T.W. Parks. Orthogonal subspace decomposition of periodic signals. In *Signals, Systems, and Computers, 1999. Conference Record of the Thirty-Third Asilomar Conference on*, volume 2, pages 1087 –1091 vol.2, 1999.

[89] S. Nagaoka. A short history of three chemical shifts. *Journal of Chemical Education*, 84(5):801, 2007.

[90] J. Nocedal and S. Wright. *Numerical Optimization*. Springer, 2006.

[91] D.C. Noll, D.G. Nishimura, and A. Macovski. Homodyne detection in magnetic-resonance-imaging. *Ieee Transactions On Medical Imaging*, 10(2):154–163, June 1991.

[92] D.C. Noll, J.M. Pauly, and A. Macovsky. Method of enhancing the focus of magnetic resonance images. *U.S. Patent N.US5311132*, 1994.

[93] K. Oshio and D.A. Feinberg. Grase (gradient- and spin-echo) imaging: a novel fast mri technique. *Magn Reson Med*, 20(2):344–9, Aug 1991.

[94] R. Paquin, P. Pelupessy, and G. Bodenhausen. Cross-encoded magnetic resonance imaging in inhomogeneous fields. *J Magn Reson*, 201(2):199–204, Dec 2009.

[95] D.C. Peters, M.A. Guttman, A.J. Dick, V.K. Raman, R.J. Lederman, and E.R. McVeigh. Reduced field of view and undersampled pr combined for interventional imaging of a fully dynamic field of view. *Magn Reson Med*, 51(4):761–7, Apr 2004.

[96] D.C. Peters, F.R. Korosec, T.M. Grist, W.F. Block, J.E. Holden, K.K. Vigen, and C.A. Mistretta. Undersampled projection reconstruction applied to mr angiography. *Magn Reson Med*, 43(1):91–101, Jan 2000.

[97] J.G. Pipe. Motion correction with propeller mri: application to head motion and free-breathing cardiac imaging. *Magn Reson Med*, 42(5):963–9, Nov 1999.

[98] J.G. Pipe, V.G. Farthing, and K.P. Forbes. Multishot diffusion-weighted fse using propeller mri. *Magn Reson Med*, 47(1):42–52, Jan 2002.

[99] G. Placidi. Metodi di codifica e decodifica dei segnali e ricostruzione di immagini nel dominio delle accelerazioni spaziali per imaging e spettroscopia di risonanza magnetica e relativo apparato. *Italian Patent N.AQ2007A000017*, 2007.

[100] G. Placidi. Circular acquisition to define the minimal set of projections for optimal mri reconstruction. In Reneta Barneva, Valentin Brimkov, Herbert Hauptman, Renato Natal Jorge, and João Tavares, editors, *Computational Modeling of Objects Represented in Images*, volume 6026 of *Lecture Notes in Computer Science*, pages 254–262. Springer Berlin / Heidelberg, 2010.

[101] G. Placidi. Constrained reconstruction for sparse magnetic resonance imaging. In Ratko Magjarevic, Olaf Dössel, and Wolfgang C. Schlegel, editors, *World Congress on Medical Physics and Biomedical Engineering, September 7 - 12, 2009, Munich, Germany*, volume 25/4 of *IFMBE Proceedings*, pages 89–92. Springer Berlin Heidelberg, 2010.

[102] G. Placidi, M. Alecci, S. Colacicchi, and A. Sotgiu. Fourier Reconstruction as a Valid Alternative to Filtered Back Projection in Iterative Applications: Implementation of Fourier Spectral Spatial EPR Imaging* 1. *Journal of Magnetic Resonance*, 134(2):280–286, 1998.

[103] G. Placidi, M. Alecci, and A. Sotgiu. Theory of adaptive acquisition method for image reconstruction from projections and application to epr imaging. *Journal of Magnetic Resonance, Series B*, 108(1):50–57, 1995.

[104] G. Placidi, M. Alecci, and A. Sotgiu. Angular space-domain interpolation for filtered back projection applied to regular and adaptively measured projections. *Journal of Magnetic Resonance, Series B*, 110(1):75–79, 1996.

[105] G. Placidi, M. Alecci, and A. Sotgiu. ω-space adaptive acquisition technique for magnetic resonance imaging from projections. *Journal of Magnetic Resonance*, 143(1):197–207, 2000.

[106] G. Placidi, M. Alecci, and A. Sotgiu. A general Algorithm for Magnetic Resonance Imaging Simulation: a Versatile Tool to Collect Information about Imaging Artefacts and New Acquisition Techniques. In *Health data in the information society: proceedings of MIE2002*, page 13. IOS Press, 2002.

[107] G. Placidi, D. Franchi, A. Galante, and A. Sotgiu. A novel acceleration coding/reconstruction algorithm for magnetic resonance imaging in presence of static magnetic field in-homogeneities. In *Proceedings of the 4th International Symposium on Advances in Visual Computing, Part II*, ISVC '08, pages 1115–1124, Berlin, Heidelberg, 2008. Springer-Verlag.

[108] G. Placidi, D. Franchi, and A. Maurizi. Characterisation of a coding/reconstruction algorithm on mri simulated noisy data. *Medical Measurement and Applications*, 0:156–159, 2009.

[109] G. Placidi, D. Franchi, A. Maurizi, and A. Sotgiu. Recent patents on magnetic resonance imaging sequences in presence of static magnetic field in-homogeneity. *Recent Patents on Biomedical Engineering*, 2(1):73–80, 2009.

[110] G. Placidi and A. Sotgiu. A novel algorithm for the reduction of under-sampling artefacts in magnetic resonance images. *Magnetic Resonance Imaging*, 22(9):1279–1287, 2004.

[111] K.P. Pruessmann, M. Weiger, M.B. Scheidegger, and P. Boesiger. Sense: sensitivity encoding for fast mri. *Magn Reson Med*, 42(5):952–62, Nov 1999.

[112] E.M. Purcell, H.C. Torrey, and R.V. Pound. Resonance absorption by nuclear magnetic moments in a solid. *Physical Review*, 69(1-2):37–38, 1946.

[113] X. Qu, X. Cao, D. Guo, C. Hu, and Z. Chen. Combined sparsifying transforms for compressed sensing mri. *Electronics Letters*, 46(2):121 –123, 21 2010.

[114] X. Qu, X. Cao, D. Guo, C. Hu, and Z. Chen. Compressed sensing mri with combined sparsifying transforms and smoothed l0 norm minimization. In *Acoustics Speech and Signal Processing (ICASSP), 2010 IEEE International Conference on*, pages 626 –629, march 2010.

[115] V. Rasche, D. Holz, and R. Proksa. Mr fluoroscopy using projection reconstruction multi-gradient-echo (prmge) mri. *Magn Reson Med*, 42(2):324–34, Aug 1999.

[116] T.G. Reese, T.L. Davis, and R.M. Weisskoff. Automated shimming at 1.5 t using echo-planar image frequency maps. *J Magn Reson Imaging*, 5(6):739–45, 1995.

[117] B.R. Rosen, J.W. Belliveau, and D. Chien. Perfusion imaging by nuclear magnetic resonance. *Magn Reson Q*, 5(4):263–81, Oct 1989.

[118] V.M. Runge, J.F. Timoney, and N.M. Williams. Magnetic resonance imaging of experimental pyelonephritis in rabbits. *Invest Radiol*, 32(11):696–704, Nov 1997.

[119] K. Scheffler and J. Hennig. Reduced circular field-of-view imaging. In *Book of abstracts: Sixth Annual Scientific Meeting and Exhibition. ISMRM*, page 180, 1998.

[120] J.F. Schenck. The role of magnetic susceptibility in magnetic resonance imaging: MRI magnetic compatibility of the first and second kinds. *MEDICAL PHYSICS-LANCASTER PA-*, 23:815–850, 1996.

[121] H. Sedarat, A.B. Kerr, J.M. Pauly, and D.G. Nishimura. Partial-fov reconstruction in dynamic spiral imaging. *Magn Reson Med*, 43(3):429–39, Mar 2000.

[122] W.A. Sethares and T.W. Staley. Periodicity transforms. *Signal Processing, IEEE Transactions on*, 47(11):2953 –2964, nov 1999.

[123] J. Shen and Y. Xiang. High fidelity magnetic resonance imaging by frequency sweep encoding and fourier decoding. *J Magn Reson*, 204(2):200–7, Jun 2010.

[124] T. Shin, J.F. Nielsen, and K.S. Nayak. Accelerating dynamic spiral mri by algebraic reconstruction from undersampled k–t space. *IEEE Trans Med Imaging*, 26(7):917–24, Jul 2007.

[125] Y. Shu, A.M. Elliott, S.J. Riederer, and M.A. Bernstein. 3d ringlet: spherical shells trajectory for self-navigated 3d mri. In *Proceedings of the 13th Annual Meeting of ISMRM, Miami Beach, FL, USA, 2693, (2005)*, 2005.

[126] Y. Shu, A.M. Elliott, S.J. Riederer, and M.A. Bernstein. Motion correction properties of the shells k-space trajectory. *Magn Reson Imaging*, 24(6):739–49, Jul 2006.

[127] Y. Shu, S.J. Riederer, and M.A. Bernstein. Three-dimensional mri with an undersampled spherical shells trajectory. *Magn Reson Med*, 56(3):553–62, Sep 2006.

[128] T.B. Smith and K.S. Nayak. Mri artifacts and correction strategy. *Imaging in Medicine*, 2(4):445–457, 2010.

[129] D.K. Sodickson and W.J. Manning. Simultaneous acquisition of spatial harmonics (smash): fast imaging with radiofrequency coil arrays. *Magn Reson Med*, 38(4):591–603, Oct 1997.

[130] D.M. Spielman, J.M. Pauly, and C.H. Meyer. Magnetic resonance fluoroscopy using spirals with variable sampling densities. *Magn Reson Med*, 34(3):388–94, Sep 1995.

[131] J. Spiniak, A. Guesalaga, R. Mir, M. Guarini, and P. Irarrazaval. Undersampling k-space using fast progressive 3d trajectories. *Magn Reson Med*, 54(4):886–92, Oct 2005.

[132] M.R. Terk, H.E. Simon, R.C. Udkoff, and P.M. Colletti. Halfscan: Clinical applications in mr imaging. *Magnetic Resonance Imaging*, 9(4):477 – 483, 1991.

[133] the mathworks. http://www.mathworks.com.

[134] J. Tintera, G. Schaub, J. Gawehn, and S. Stoeter. Functional mri with keyhole technique. In *Proc., 1st International Conference on Functional Mapping of the Human Brain, Paris, 1995*, page 124, 1995.

[135] S.I. Tomonaga. *The Story of Spin*. University of Chicago Press Ltd, London, 1997.

[136] C.M. Tsai and D.G. Nishimura. Reduced aliasing artifacts using variable-density k-space sampling trajectories. *Magn Reson Med*, 43(3):452–8, Mar 2000.

[137] J. Tsao. On the unfold method. *Magn Reson Med*, 47(1):202–7, Jan 2002.

[138] J. Tsao, P. Boesiger, and K.P. Pruessmann. k-t blast and k-t sense: dynamic mri with high frame rate exploiting spatiotemporal correlations. *Magn Reson Med*, 50(5):1031–42, Nov 2003.

[139] R. Van de Walle, I. Lemahieu, and E. Achten. Magnetic resonance imaging and the reduction of motion artifacts: review of the principles. *Technol Health Care*, 5(6):419–35, Dec 1997.

[140] R.L. Van Metter, J. Beutel, and H. Kundel. *Handbook of medical imaging. Volume 1 Physics and psychophysics*. SPIE Press, Bellingham, WA, 2000.

[141] J.J. van Vaals, M.E. Brummer, W.T. Dixon, H.H. Tuithof, H. Engels, R.C. Nelson, B.M. Gerety, J.L. Chezmar, and J.A. den Boer. "keyhole" method for accelerating imaging of contrast agent uptake. *J Magn Reson Imaging*, 3(4):671–5, 1993.

[142] J.J. van Vaals, G.H. van Yperen, and R.W. de Boer. Real-time mr imaging using the lolo (local look) method for interactive and interventional mr at 0.5t and 1.5t. In *Book of abstracts: Second Annual Meeting of the Society of Magnetic Resonance Imaging. SMRI*, page 421, 1994.

[143] K.K. Vigen, D.C. Peters, T.M. Grist, W.F. Block, and C.A. Mistretta. Undersampled projection-reconstruction imaging for time-resolved contrast-enhanced imaging. *Magn Reson Med*, 43(2):170–6, Feb 2000.

[144] A.G. Webb, Z.P. Liang, R.L. Magin, and P.C. Lauterbur. Applications of reduced-encoding mr imaging with generalized-series reconstruction (rigr). *J Magn Reson Imaging*, 3(6):925–8, 1993.

[145] S. Weiss and V. Rasche. Projection-reconstruction reduced [correction of reduces] fov imaging. *Magn Reson Imaging*, 17(4):517–25, May 1999.

[146] E.B. Welch, A. Manduca, R.C. Grimm, H.A. Ward, and C.R. Jr Jack. Spherical navigator echoes for full 3d rigid body motion measurement in mri. *Magn Reson Med*, 47(1):32–41, Jan 2002.

[147] N. Wilke, C. Simm, J. Zhang, J. Ellermann, X. Ya, H. Merkle, G. Path, H. Lüdemann, R.J. Bache, and K. Uğurbil. Contrast-enhanced first pass myocardial perfusion imaging: correlation between myocardial blood flow in dogs at rest and during hyperemia. *Magn Reson Med*, 29(4):485–97, Apr 1993.

[148] A.H. Wilman, S.J. Riederer, B.F. King, J.P. Debbins, P.J. Rossman, and R.L. Ehman. Fluoroscopically triggered contrast-enhanced three-dimensional mr angiography with elliptical centric view order: application to the renal arteries. *Radiology*, 205(1):137–46, Oct 1997.

[149] J. L Wilson, M. Jenkinson, I. de Araujo, M.L. Kringelbach, E.T. Rolls, and P. Jezzard. Fast, fully automated global and local magnetic field optimization for fmri of the human brain. *Neuroimage*, 17(2):967–76, Oct 2002.

[150] S.T. Wong and M.S. Roos. A strategy for sampling on a sphere applied to 3d selective rf pulse design. *Magn Reson Med*, 32(6):778–84, Dec 1994.

[151] Q.S. Xiang and R.M. Henkelman. K-space description for mr imaging of dynamic objects. *Magn Reson Med*, 29(3):422–8, Mar 1993.

[152] J. Xiong, P.T. Fox, and J.H. Gao. The effects of k-space data undersampling and discontinuities in keyhole functional mri. *Magn Reson Imaging*, 17(1):109–19, Jan 1999.

[153] Q.X. Yang and M.B. Smith. Mri with removal of field inhomogeneity artefacts. *World patent N.WO9930187*, 1999.

[154] C. Yao, J.D. Hale, L.E. Crooks, and L. Kaufman. Mri compensated for spurious nmr frequency/phase shifts caused by spurious changes in magnetic fields during nmr data measurement processes. *U.S. Patent N.US4885542*, 1989.

[155] H. Yu and G. Wang. Compressed sensing based interior tomography. *Phys Med Biol*, 54(9):2791–805, May 2009.

[156] C. Zang-Hee, J.R. Wong, and K. Edward. Fringe field mri. *U.S. Patent N.US5023554*, 1991.

Index

T - #0428 - 071024 - C19 - 234/156/10 - PB - 9780367381431 - Gloss Lamination